Solar Power For Beginners Bible 2024

10 Books in 1 Your Comprehensive Guide to Mastering Solar Energy from Basics to Off-grid Living, Urban Solutions, and Tomorrow's Policies

By Thomas Daughtler

Table Of Contents

BOOK 1
The Solar Revolution

CHAPTER 1: History of Solar Energy

The Ancient Solar Innovations

Before delving into the intricacies of modern-day solar energy, it's vital to look back and understand how ancient civilizations utilized the sun. The story of solar innovations isn't one of mere decades or even centuries—it spans millennia, revealing the deeply entrenched relationship between humanity and the sun.

Harnessing the Sun's Warmth: The Ancestral Homes

When we speak of ancient solar innovations, it's not about solar panels or photovoltaic cells. It's about understanding the sun's trajectory, its heat, and its light, then adapting architectural designs to capitalize on its gifts. Take, for instance, the indigenous tribes of North America. Without complex tools or advanced knowledge, they constructed their homes in orientations that maximized sun exposure during winter while minimizing it during summer, a practice evident in the Puebloan structures of the American Southwest. These homes, built into cliff faces or as compact villages, utilized the sun's heat for both warmth and a way to regulate indoor temperatures.

In ancient Greece, similar concepts were embraced. City planning, especially in regions with cooler climates, involved the strategic orientation of homes to capture the most sunlight. It ensured not only that the homes stayed warm but that their interiors were well-lit, reducing the reliance on candles or oil lamps.

Architectural Marvels: The Sun Temples

Temples dedicated to sun gods, such as those in Egypt for the god Ra or in India for the deity Surya, weren't merely places of worship. Their very design was a nod to the power of the sun. Built with precise alignments, these temples would illuminate divine statues or sacred spaces during particular solar events, like solstices or equinoxes. It demonstrated an understanding of solar patterns, an early form of astronomy, intertwined with spiritual beliefs.

The famous temple of Abu Simbel in Egypt stands as a testament to this. Its inner sanctum lights up twice a year, aligning perfectly with the birth and coronation dates of Pharaoh Ramses II. This isn't mere coincidence—it's calculated architecture.

Reflecting Brilliance: Using the Sun for Signaling and Defense

While most solar innovations were about harnessing warmth or light, there were other ingenious uses. The Greeks, ever the pioneers in creative thinking, devised a weapon called the "Archimedes Heat Ray." As legends have it, this weapon was used during the Siege of Syracuse. Using large polished shields or mirrors, soldiers focused sunlight onto the approaching enemy ships, setting them ablaze. While the historical accuracy of this event is debated, it showcases early ideas of using sunlight as a concentrated force.

Similarly, cultures around the world used polished surfaces for signaling, harnessing sunlight to send messages across vast distances long before the advent of telecommunications.

A Natural Clock: Understanding Time with the Sun

Timekeeping, an essential facet of any organized society, relied heavily on solar movements. Sundials, used across different ancient civilizations from Rome to China, showcased the ingenuity of tracking time using the sun's shadows. These weren't mere stone structures but intricately designed pieces, often decorated and aligned with astrological beliefs.

The obelisks of ancient Egypt served dual purposes. Apart from their religious significance, they acted as giant sundials. Their shadows, moving with the sun, would indicate different times of the day, helping people schedule their activities, from prayer rituals to market hours.

Modern Solar Breakthroughs

While ancient civilizations laid the groundwork by observing and incorporating the sun's patterns into their daily lives, the modern era marked a shift. Humanity began to actively harness the sun's energy for more direct uses. This chapter uncovers the milestones that shaped the solar industry, catapulting it from humble beginnings to the forefront of the global energy revolution.

A Spark in the 19th Century: The Birth of Photovoltaic Science

The 1800s were a time of rapid industrialization and scientific inquiry. Amidst this backdrop, in 1839, a French physicist named Alexandre-Edmond Becquerel made a groundbreaking discovery. While experimenting with an electrolytic cell, he noticed that when it was exposed to sunlight, it generated more electricity. This phenomenon, now known as the photovoltaic effect, was the first step towards understanding how sunlight could be directly converted into electricity.

However, it would take over half a century before the first genuine solar cell was created. In 1883, American inventor Charles Fritts crafted a solar cell using selenium, a chemical element. Though it was rudimentary and had a minuscule efficiency, it set the stage for what was to come.

Silicon Takes the Stage: The Rise of Efficient Solar Panels

Jump to the 20th century, and the world witnessed an acceleration in solar innovation. The 1950s, in particular, were transformative. In 1954, Bell Labs introduced the first silicon solar cell. Unlike its selenium predecessor, this solar cell boasted an efficiency rate of about 6%, a significant leap that opened the doors to practical applications.

Soon, satellites in the burgeoning space industry adopted these solar cells, marking the beginning of solar technology's integration into mainstream applications.

A Sustainable Dream: The Energy Crisis and Solar's Solution

The global energy crisis of the 1970s, spurred by political and economic turmoil, forced nations to reconsider their heavy reliance on fossil fuels. With soaring oil prices and an uncertain energy future, there was a renewed interest in alternative energy sources, and solar was right at the forefront.

Governments and private sectors began investing heavily in solar research, leading to technological advancements and reductions in production costs. Photovoltaic technology improved, efficiencies increased, and solar panels became more affordable for the average consumer.

Into the 21st Century: Solar Goes Mainstream

The momentum gained in the 20th century carried well into the 21st. As the realities of climate change became increasingly evident, the shift towards sustainable energy solutions became not just desirable but imperative. Solar energy, with its promise of clean and infinite power, was an obvious choice.

Innovations in thin-film technology, concentrating solar power systems, and solar storage solutions further propelled the industry forward. As of the 2010s, many countries, realizing the vast potential and environmental benefits of solar, implemented subsidies and incentives to promote its adoption.

Not Just Panels: Modern Applications of Solar Energy

While solar panels on rooftops became a common sight, the modern era also saw diverse applications of solar energy. From solar-powered cars participating in international challenges across the Australian outback to solar water heaters providing sustainable solutions for home heating needs, the applications seemed endless.

In cities, solar street lights became increasingly popular, providing illumination without the need for grid connectivity. In remote areas, where electricity lines couldn't reach, solar mini-grids brought power, transforming lives and empowering communities.

From Becquerel's early observations to the sprawling solar farms of today that power entire cities, the journey of modern solar breakthroughs is nothing short of spectacular. These innovations aren't just technical achievements; they represent hope—a vision of a world where energy is clean, abundant, and harmonious with nature.

CHAPTER 2: Benefits of Solar Power

Environmental Advantages

In an era where urban landscapes are often clouded by smog and industrial outputs, the demand for cleaner solutions becomes paramount. Solar energy stands as one of the most impactful and immediate answers to this call. Unlike traditional fossil fuels, solar power generation doesn't release harmful pollutants like carbon dioxide, nitrogen oxides, and particulate matter into the atmosphere. By diminishing our reliance on coal, oil, and natural gas, solar power plays a pivotal role in reducing air pollution, improving public health, and preserving the environment for future generations.

Guardians of the Atmosphere: Slowing Climate Change

Perhaps one of the most urgent environmental crises of our time is climate change. As the Earth's temperature continues to rise due to the greenhouse effect, we witness more frequent and severe weather events, rising sea levels, and a disruption of natural ecosystems. Solar energy, by its nature, emits zero greenhouse gases once installed. Every kilowatt-hour generated by the sun instead of fossil fuels prevents a significant amount of carbon emissions from entering our atmosphere, making solar power a primary weapon in the battle against global warming.

Protecting Precious Resources: Water Conservation

Water is one of our planet's most valuable resources. Traditional power generation methods, particularly those using coal or nuclear fuel, require vast amounts of water for cooling. These methods can lead to water pollution, ecosystem damage, and increased competition for water resources in already water-scarce regions. Solar photovoltaic cells, on the other hand, require no water to generate electricity. By adopting solar energy, we not only conserve water but also alleviate the stress on local water resources.

A Land of Harmony: Preserving Habitats and Biodiversity

While solar farms require land, their environmental footprint is significantly less damaging than other forms of power generation. Unlike mining operations or hydroelectric dams, which can devastate local habitats, solar installations can be designed to coexist harmoniously with the local environment. There are numerous examples of multi-use solar farms where agricultural activities or wildlife conservation occur simultaneously with energy production. By choosing solar, we're making a commitment to preserve the intricate tapestry of life that our ecosystems hold.

Echoes of a Clean Future: Reducing Noise Pollution

While often overlooked, noise pollution is a tangible concern, especially in densely populated areas. Traditional power plants and generators produce significant noise, which can be disruptive to both humans and wildlife. Solar installations, in contrast, operate in near silence. By embracing solar, we contribute to quieter, more serene urban and rural landscapes.

Economic Savings

At the dawn of the solar revolution, the initial costs associated with solar installations were notably high, making it a less attractive option for the average consumer. However, as technology has progressed and the solar industry matured, costs have plummeted. Today, harnessing the power of the sun is not only an environmentally conscious decision but also a highly cost-effective one.

The Decline of Installation Costs

One of the most profound changes in the solar industry has been the significant decrease in the costs of solar panels and associated equipment. Innovations in manufacturing, global supply chains, and increased competition have driven down prices, making solar installations more affordable for homeowners, businesses, and utilities alike. This trend shows no sign of slowing down, ensuring that solar remains an economically viable choice for many years to come.

The Beauty of Self-sufficiency

Imagine a future where monthly utility bills are a thing of the past. With solar energy, this dream is closer to reality than one might think. By generating your own electricity, you not only become less dependent on the grid but also protect yourself from the ever-increasing energy prices. This level of self-sufficiency offers an unmatched level of economic freedom.

Incentives, Rebates, and Tax Credits

Governments around the world recognize the importance of transitioning to renewable energy. As a result, many offer attractive incentives, rebates, and tax credits to those who choose to invest in solar power. These financial instruments can significantly offset installation costs and hasten the return on investment, making solar an even more appealing choice for the economically-minded individual.

Maintenance: The Low-cost Promise

Unlike many other forms of electricity generation, solar power systems require minimal maintenance. Once installed, solar panels can last for 25 years or more with only occasional cleaning and periodic checks. This longevity, combined with the low upkeep costs, ensures that the economic benefits of solar extend well beyond the initial savings on electricity bills.

Boosting Property Value: A Bright Investment

As awareness of environmental issues grows, properties equipped with solar power systems are becoming increasingly desirable. Numerous studies have shown that homes with solar installations sell faster and at higher prices than those without. Investing in solar isn't just about saving on monthly bills; it's about enhancing the overall value of one's property.

Job Creation and Economic Growth

On a broader scale, the rise of solar energy has led to significant job creation in manufacturing, installation, maintenance, and research sectors. This not only provides economic opportunities for countless individuals but also stimulates local economies. The solar industry's growth promises a brighter economic future for communities around the globe.

CHAPTER 3: Types of Solar Systems

Grid-tied vs. Off-grid

For many, the idea of using solar power conjures up images of being entirely off this grid, living in self-sufficient harmony. While this is a viable and appealing approach for some, many opt for a more interconnected solution: the grid-tied solar system.

The Basics of Grid-tied Systems

A grid-tied solar system, as its name suggests, is directly connected to the local electrical grid. During the day, when the sun is shining, solar panels generate electricity. Any excess electricity that isn't used by the home or business is then fed back into the grid. Conversely, during cloudy days or at night, when the solar panels aren't producing enough electricity to meet demands, power can be drawn from the grid to fill in the gaps.

The Financial Benefits of Feeding the Grid

One of the major advantages of a grid-tied system is the potential for net metering. Depending on local regulations, homeowners can receive credits for the excess power they contribute to the grid. This can lead to significantly reduced – or even nonexistent – electricity bills. In some cases, utilities might even pay the homeowner for the surplus power, turning their solar installation into a revenue-generating asset.

Streamlined Systems for the Modern Homeowner

For those who prioritize ease and simplicity, grid-tied systems often require less equipment and maintenance than their off-grid counterparts. There's no need for extensive battery storage systems, given the reliance on the grid during downtimes. This can make grid-tied solutions both more affordable and easier to integrate into existing homes and buildings.

The Off-grid Dream: A World Untethered

While the convenience of grid-tied systems appeals to many, there's an undeniable allure to the idea of complete energy independence. This is where off-grid solar systems come into play, offering a way to break free from the confines of traditional energy infrastructure.

Understanding Off-grid Systems

An off-grid solar system operates independently from the local electricity grid. All the power needed must be generated on-site, typically through a combination of solar panels and battery storage. This stored energy is crucial, providing electricity during periods when the solar panels aren't generating power.

The Freedom of Total Independence

Living off-grid offers a unique form of freedom. There's no reliance on external utilities, and homeowners are insulated from grid failures or blackouts. This independence also provides a deep sense of self-sufficiency and resilience, knowing that one's energy needs are met through their own resources.

The Challenges of Going Solo

It's essential to understand that off-grid living comes with its own set of challenges. These systems often require a more significant initial investment, especially in battery storage. They also necessitate careful energy management and planning to ensure that the stored power can meet the household's needs, especially during extended periods of cloudy weather.

Hybrid Solar Systems

While the concepts of grid-tied and off-grid solar systems each have their distinct appeals, there exists an innovative middle ground that offers homeowners the best of both worlds: the hybrid solar system. As the landscape of renewable energy evolves, hybrid systems emerge as a flexible solution that marries the reliability of grid connection with the independence and resilience of off-grid setups.

Unveiling the Hybrid Solar Approach

At its core, a hybrid solar system integrates the primary features of both grid-tied and off-grid systems. It has solar panels that produce electricity from sunlight, a connection to the main electrical grid, and, most crucially, a battery storage component. This battery storage allows for the retention of surplus energy, which can be used during times of lower solar production or during grid outages.

Advantages of Opting for Hybrid

1. **Flexibility in Energy Usage**: One of the standout features of hybrid systems is the flexibility they afford homeowners. On sun-rich days, excess power can either be stored for later use, fed back into the grid for credits (if net metering is available), or a combination of both.

2. **Resilience During Outages**: Grid failures and blackouts can be more than just an inconvenience; they can disrupt daily life and even pose safety risks. With a hybrid system, the stored energy in the batteries can be harnessed during such times, ensuring continued power supply.

3. **Economic Benefits**: Hybrid systems, when paired with net metering policies, can offer substantial economic benefits. Not only can homeowners reduce their electricity bills, but they can also earn credits or even income from feeding surplus energy back into the grid.

4. **Optimal Energy Management**: With smart controllers and modern inverter technology, hybrid systems can be programmed to use stored energy during peak pricing periods, further reducing electricity costs.

Considerations When Going Hybrid

While the benefits of hybrid systems are numerous, it's essential to approach them with a holistic perspective. The initial setup cost, particularly for high-quality battery storage, can be higher than a simple grid-tied system. Maintenance, while not overly complex, does require attention to both the grid-connected components and the battery storage. Finally, as with any solar setup, the system's efficiency is contingent upon the quality of the installation, the equipment used, and the location's solar irradiance.

A Vision of the Future

Hybrid solar systems represent an evolutionary step in the renewable energy domain. They encapsulate the vision of a world where homeowners can enjoy the reliability and economic advantages of grid connection while simultaneously harnessing the empowerment and resilience of off-grid capabilities. As technology continues to advance, it's likely that the adoption of hybrid solutions will grow, making them a cornerstone of the global shift towards sustainable living.

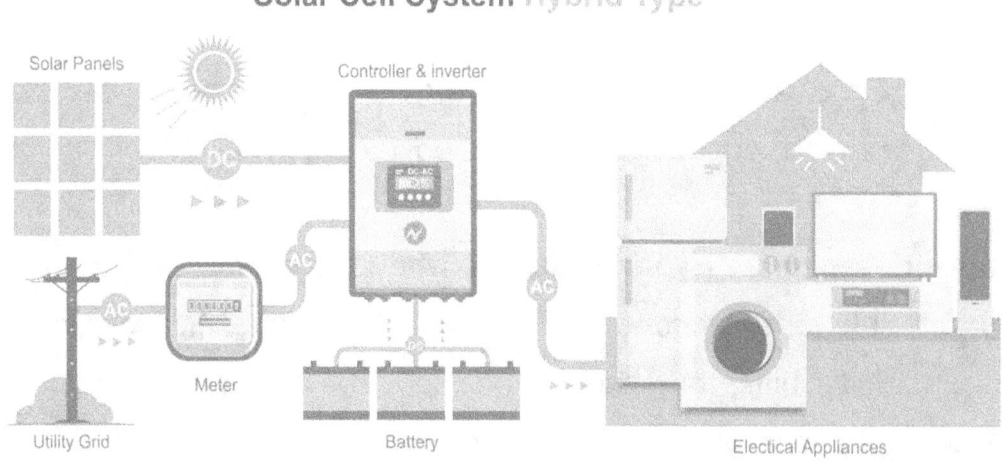

Solar Cell System Hybrid Type

CHAPTER 4: Solar Energy Around the World

Leaders in Solar Implementation

The sun, a consistent and potent force of nature, has been acknowledged and revered by many ancient civilizations. In modern times, its immense power has been harnessed to address one of the most pressing challenges of our era: sustainable energy production. As concerns over climate change, depleting fossil fuels, and energy security mount, nations worldwide have turned their gaze towards the sun, seeking solutions that are both environmentally and economically viable.

This global shift towards solar energy has seen several countries emerge as frontrunners, leading by example in solar implementation, technological innovation, and policy framework. These nations are not just advocates for renewable energy; they embody the vision of a sustainable future, showcasing the transformative power of solar energy on a national scale.

Germany: Pioneering the Solar Movement

Known for its engineering marvels and a long-standing commitment to sustainability, Germany has consistently been at the forefront of the solar revolution. Despite its relatively overcast climate, the nation's solar capacity is impressive, driven by a potent combination of forward-thinking policies, public enthusiasm, and technological prowess.

Germany's *Energiewende* or "energy transition" policy, implemented in the early 2000s, aimed at dramatically increasing renewable energy adoption. This aggressive shift towards renewables, combined with a favorable feed-in tariff system, galvanized both individual homeowners and businesses to invest in solar installations.

But beyond policy and capacity, Germany's role as a leader in solar implementation is cemented by its emphasis on research, innovation, and education. The nation hosts numerous international solar conferences, fostering a global exchange of ideas and best practices, ensuring that the solar torch burns bright for generations to come.

China: The Solar Powerhouse

If Germany is recognized for pioneering the solar movement, China stands out for its sheer scale of implementation. Over the past decade, China has transformed from a country with minimal solar infrastructure to the world's largest producer of photovoltaic panels and a global leader in solar capacity.

Several factors have propelled China's meteoric rise in the solar domain. Firstly, the government's ambitious renewable energy targets, paired with significant financial incentives, have stimulated both domestic demand and international exports. Secondly, Chinese manufacturers have achieved significant economies of scale, driving down the global price of solar panels and making solar adoption more accessible worldwide.

However, China's solar journey is not just about vast solar farms and mass production. There's a conscious effort to integrate solar installations into urban landscapes seamlessly, with innovations like solar highways and floating solar farms becoming symbols of the nation's renewable ambitions.

India: A Sunlit Promise

India, with its abundant sunshine and rapidly growing energy needs, presents a unique canvas for solar implementation. The nation's commitment to solar is not merely a nod to sustainable practices but a vital strategy to address energy security, reduce carbon emissions, and provide electricity to remote areas.

The Indian government's ambitious target of achieving 100 GW of solar capacity by 2022 speaks volumes about its solar aspirations. The establishment of the International Solar Alliance (ISA) in partnership with France further underlines India's intent to spearhead global solar collaboration, especially among sun-rich equatorial nations.

From vast solar parks in its western deserts to innovative canal-top solar installations that prevent water evaporation, India's approach to solar is diverse, adaptive, and deeply rooted in its socio-economic fabric.

The United States: A Mosaic of Solar Endeavors

The story of solar in the United States is multifaceted, reflecting the nation's vast geographical and cultural expanse. While federal initiatives like the Solar Investment Tax Credit (ITC) have provided a nationwide impetus, the real momentum in the U.S. solar movement often stems from state-level initiatives and private enterprise.

States like California, with its progressive renewable mandates, and Texas, leveraging its vast sun-soaked lands, have become regional leaders in solar implementation. Additionally, American corporations, universities, and non-profits are increasingly investing in solar, both for its economic benefits and as a commitment to sustainability.

Emerging Solar Markets

Solar energy's evolution is akin to the lifecycle of dawn. While established markets like Germany, China, and the U.S. have seen the sun reach its zenith, many nations are currently experiencing the initial golden hues of a solar dawn. These emerging markets, with their burgeoning solar ambitions, are crucial in realizing the vision of a sustainable global energy landscape.

Africa: Illuminating the Dark Continent

The African continent, with its vast landmass bathed in sunlight, holds immense solar potential. For years, large parts of Africa remained in the dark, with millions lacking access to basic electricity. However, the solar revolution is changing this narrative.

Nations like Morocco, with its colossal Noor Ouarzazate solar complex, have showcased the continent's solar capabilities. In sub-Saharan Africa, smaller scale, off-grid solar solutions are transforming rural landscapes. Portable solar kits and mini-grids are empowering communities, allowing kids to study after dusk, farmers to increase productivity, and small businesses to thrive.

Southeast Asia: Harnessing the Tropical Sun

Southeast Asia, a region characterized by its tropical climate and growing energy demands, presents a compelling case for solar adoption. Countries like Vietnam have made remarkable strides in a short span. In the space of a few years, Vietnam transitioned from a negligible solar market to one of the largest in Southeast Asia, aided by favorable policies and the decreasing costs of photovoltaic technology.

The Philippines, with its archipelagic geography, is leveraging solar to electrify remote islands, reducing dependence on expensive diesel generators. Meanwhile, Malaysia, through its Large Scale Solar (LSS) program, is augmenting its national grid with significant solar input.

Latin America: From the Andes to the Amazon

The vast landscapes of Latin America, from the towering Andes to the sprawling Amazon, are increasingly dotted with solar panels. Countries like Brazil, with vast tracts of sunlit savannahs, and Chile, with its sun-drenched Atacama desert, are natural candidates for solar domination.

Chile's commitment to solar is evident in its thriving solar farms, supplying power not just for domestic use but also for energy-intensive industries like mining. Argentina, through its RenovAr program, aims to generate a significant portion of its electricity from renewable sources, with solar playing a pivotal role.

Middle East: Oil to Solar – A Paradigm Shift

The Middle East, the epicenter of the global oil industry, is witnessing a paradigm shift. As the world moves away from fossil fuels, nations like Saudi Arabia, the UAE, and Jordan are investing heavily in solar infrastructure. The symbolism is palpable - vast oil fields are slowly being neighbored by sprawling solar farms.

Dubai's Mohammed bin Rashid Al Maktoum Solar Park is a testament to the region's solar ambitions. Combining photovoltaic and concentrated solar power (CSP) technologies, the park aims to produce thousands of megawatts of electricity, making it one of the largest solar installations globally.

The Momentum of Change

The transition to solar in these emerging markets is not just about electricity generation. It's about empowerment, progress, and challenging the status quo. It's about young entrepreneurs in Nairobi selling solar lanterns, researchers in Bangkok innovating in solar storage, and indigenous communities in the Amazon using solar to pump clean water.

While challenges persist, from financing hurdles to technological barriers, the momentum in these emerging solar markets is undeniable. With every installed panel, a clearer, brighter, and more sustainable path is illuminated for these nations.

BOOK 2

Solar Fundamentals For Beginners

CHAPTER 1: Introduction to Photovoltaics

How Solar Cells Work

The core essence of our universe is energy, with the sun serving as a colossal powerhouse, dispensing endless streams of radiant energy every second. But, how do we, on our tiny blue planet, harness this energy? How do we transform the sun's beaming rays into power that charges our devices, lights our homes, and runs our industries? Enter the magical world of photovoltaics.

A Brief Dip into Quantum Physics

Solar cells, the fundamental units of a solar panel, rely on quantum physics. At a granular level, everything around us is made of atoms, with a nucleus at the center and electrons revolving in specific orbits. When sunlight, composed of tiny energy packets called photons, hits these electrons, a fascinating phenomenon occurs.

Photons carry energy corresponding to their wavelength. Blue and violet photons carry more energy than their red and yellow counterparts. When these high-energy photons strike the electrons of a solar cell, they transfer their energy. This extra energy can, sometimes, eject the electron from its standard position, leading to what we refer to as the 'photoelectric effect.'

Creating an Electric Field

Solar cells are crafted from silicon, a semiconductor. To understand the working of a solar cell, one needs to comprehend the nature of semiconductors. Semiconductors can be manipulated to enhance their properties. In the context of solar cells, silicon is typically combined with other materials, creating positive (p-type) and negative (n-type) layers.

The intersection of these layers is called the 'p-n junction.' At this junction, electrons from the n-side move to the p-side, creating an imbalance. This movement establishes an electric field. The importance of this field? It's the driving force that pushes electrons, dislodged by photons, in a specific direction, setting the stage for the creation of an electric current.

From Displaced Electrons to Usable Power

When sunlight hits a solar cell, the photoelectric effect displaces numerous electrons. However, due to the electric field at the p-n junction, these electrons are forced to move in a particular direction, creating a flow. This movement of electrons is what we refer to as electric current.

The solar cell has conductive plates on its sides to collect these electrons and transfer them to wires, thus providing power in a usable form. This is the fundamental principle behind solar energy conversion. When we connect multiple solar cells, we get a solar panel, capable of generating a substantial amount of power.

Factors Affecting Efficiency

While the concept seems straightforward, in reality, not all incoming sunlight gets converted into electricity. The efficiency of a solar cell, which denotes the portion of sunlight's energy that can be transformed into electric power, depends on various factors.

1. **Material Quality:** The purity of silicon and the quality of materials used can significantly impact efficiency. Monocrystalline silicon cells, for instance, have a higher efficiency than their polycrystalline counterparts due to their superior purity.

2. **Photon Energy:** As discussed earlier, not all photons have the same energy. Some might have too little energy to displace an electron, while others might have excess energy, which unfortunately gets wasted as heat.

3. **Cell Temperature:** Solar cells dislike excessive heat. The hotter they get, the lesser the electricity they produce from a given amount of sunlight.

4. **Angle and Intensity of Sunlight:** The angle at which sunlight hits the cell and its intensity can also play a vital role. This is why solar tracking systems, which adjust the panel's angle to face the sun directly, can enhance efficiency.

A Future Filled with Possibilities

As research deepens, scientists are exploring ways to enhance efficiency, make solar cells from alternative materials, and reduce production costs. Concepts like tandem solar cells, which layer materials to absorb a broader spectrum of sunlight, are pushing the boundaries of what's possible.

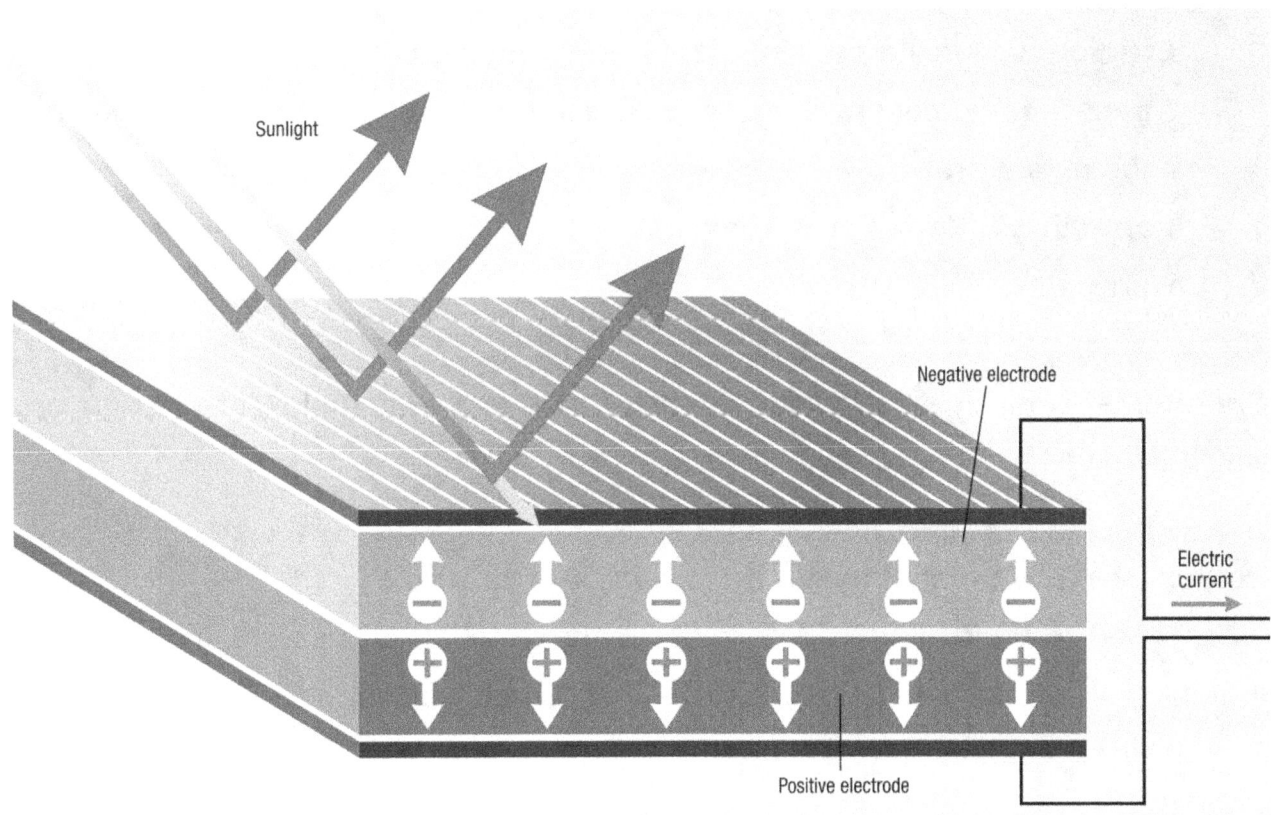

Types of Solar Panels

Solar panels, while rooted in the foundational principle of converting sunlight into electricity, have branched out into various types, each with its unique strengths, efficiencies, and applications. Let's traverse this exciting array of solar panel choices.

Monocrystalline Solar Panels: The Pure Breed

The name "monocrystalline" is derived from the method of their production. These panels are made from a single, pure crystal of silicon. Visualize it as a continuous crystal lattice with a uniform look, typically a dark black color.

Strengths:

- **Efficiency:** Among the highest, usually ranging between 15-20%.
- **Space:** They generate more electricity per square foot, ideal for space-constrained areas.
- **Longevity:** Known for their durability, they often come with extended warranties.

Considerations: They tend to be pricier due to the complex production process which involves growing a single crystal silicon.

Polycrystalline Solar Panels: A Mosaic of Crystals

As opposed to their mono counterparts, polycrystalline panels are made by melting multiple silicon fragments together. They're recognized by their blue, speckled appearance, resulting from the multiple crystals reflecting light.

Strengths:

- **Cost:** Generally more affordable as the production process is simpler and less wasteful.

- **Versatility:** Suitable for a range of applications, especially where budget is a concern.

Considerations: Their efficiency is marginally lower, typically around 13-18%. Their lifespan might also be a tad shorter than monocrystalline panels.

Thin-Film Solar Panels: Sleek and Flexible

Stepping away from the traditional crystalline structure, thin-film panels are produced by placing one or more films of photovoltaic material onto a substrate. They're characterized by their thin, sleek profile.

Strengths:

- **Flexibility:** Some types can be flexible, allowing for varied applications, including curved surfaces.
- **Aesthetics:** Their uniform appearance can be more visually appealing for certain installations.
- **Performance:** They perform relatively well in low-light conditions.

Considerations: Lower efficiency (typically 10-12%) means they require more space. Their lifespan is generally shorter than crystalline panels.

BIPV (Building Integrated Photovoltaics): Architecture Meets Solar

Blending solar technology with building architecture, BIPV panels are not just functional but form part of the building's design, such as rooftops, facades, or even windows.

Strengths:

- **Aesthetics:** Seamlessly integrates with the building design.

- **Dual-Purpose:** Serves architectural and energy-generating purposes simultaneously.

Considerations: Costs can be higher due to custom designs and installations. Efficiency varies based on design and application.

Concentrated PV Cell (CVP): Intensity Amplified

These panels are akin to a magnifying glass for sunlight. Equipped with curved mirror panels and lenses, they focus sunlight onto a small, highly efficient PV cell.

Strengths:

- **High Efficiency:** Can achieve efficiencies above 40% under optimal conditions.
- **Tracking Systems:** Often combined with solar tracking systems to keep the panel oriented towards the sun.

Considerations: Typically used in solar farms and not residential setups. They require direct sunlight and might not perform as well on cloudy days.

The diversity in solar panels is a testament to the adaptability of solar technology. While the fundamental principle remains consistent – converting sunlight into electricity – the means and methods have diversified to cater to varied needs, environments, and aspirations.

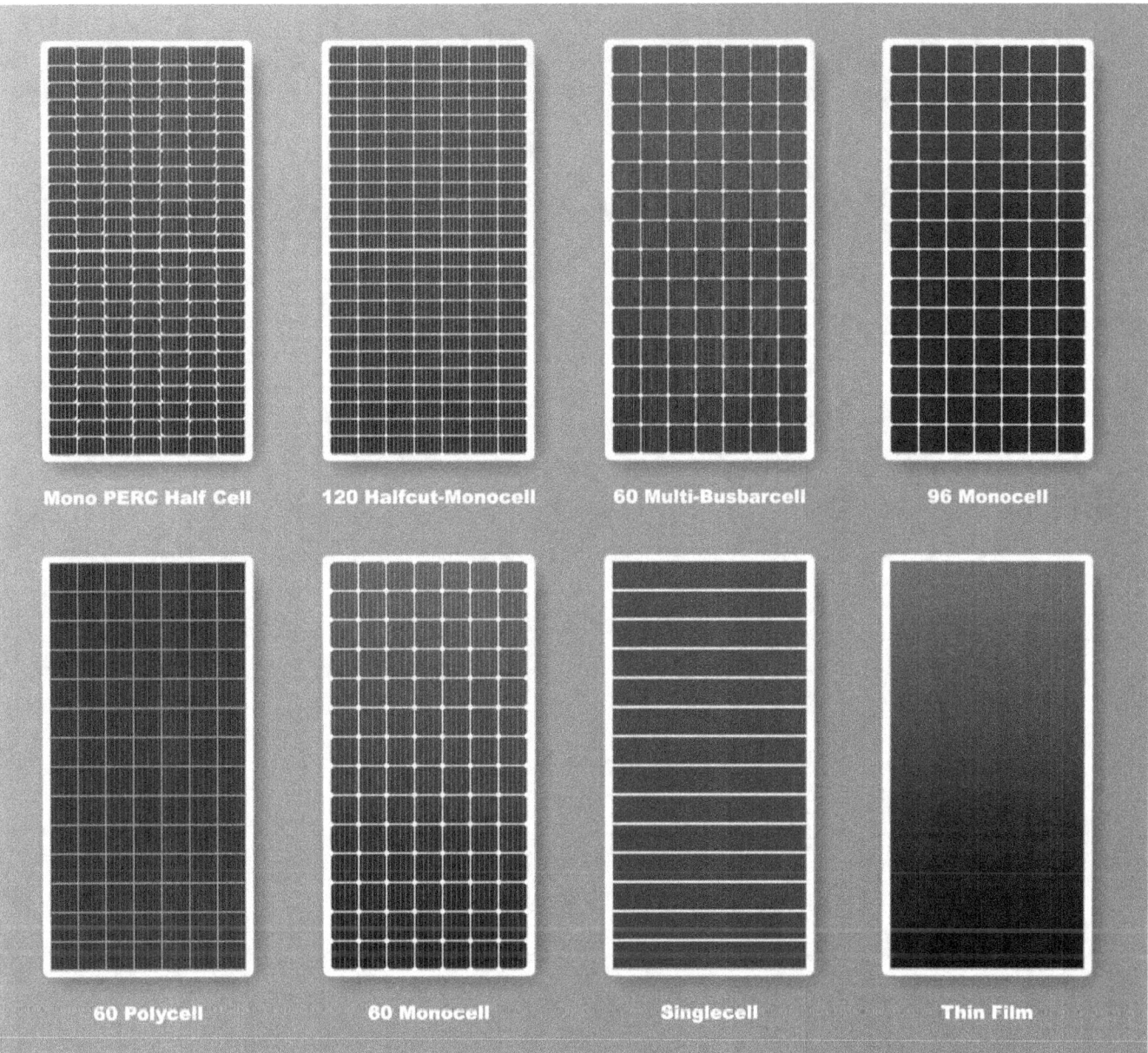

Mono PERC Half Cell   120 Halfcut-Monocell   60 Multi-Busbarcell   96 Monocell

60 Polycell   60 Monocell   Singlecell   Thin Film

CHAPTER 2: Solar Power System Components

Inverters and Controllers

While solar panels might be the poster child of solar energy, the underlying components like inverters and controllers are the unsung heroes, working tirelessly behind the scenes. Let's embark on a journey to understand these critical components that ensure the smooth operation and efficiency of a solar power system.

The Role of the Inverter: From DC to AC

At its heart, the inverter's role is simple: to convert direct current (DC) produced by solar panels into alternating current (AC) that our homes and appliances use. However, this simplicity masks the critical nature of the task and the sophistication behind it.

The Two Main Types of Inverters:

1. **String Inverters:** This traditional form of inverter connects strings of solar panels together. All the DC power generated by these panels is sent to one centralized inverter, which then converts it to AC.
2. **Microinverters:** A more recent innovation, microinverters are attached to each solar panel. This decentralized approach means each panel gets its own inverter, allowing it to operate independently.

Strengths and Considerations:

- **String Inverters:** Typically less expensive and easier to maintain since there's just one central unit. However, if one panel underperforms (due to shade, debris, or damage), it can reduce the performance of the entire string.

- **Microinverters:** Offer greater efficiency as each panel operates independently. This means the system can perform optimally even if some panels are shaded or not working perfectly. The trade-off is typically a higher initial cost and more potential points of failure.

Controllers: Guardians of the Battery

While the inverter's task is to handle the current type, solar controllers (often known as charge controllers) are entrusted with managing the flow of this current to and from the batteries. They ensure batteries are charged optimally and prevent overcharging or deep discharging, both of which can reduce battery life.

Types of Solar Controllers:

1. **PWM (Pulse Width Modulation) Controllers:** They match the solar panels' voltage with the battery voltage. As the battery gets charged, the PWM narrows the amount of energy sent to the battery.
2. **MPPT (Maximum Power Point Tracking) Controllers:** These are more advanced and adjust their input to harvest the maximum power from the solar panels. They then adjust their output to suit the battery's voltage and current.

Strengths and Considerations:

- **PWM Controllers:** More affordable and suitable for smaller systems. However, they might not be as efficient, especially in cooler temperatures.
- **MPPT Controllers:** Can boost system efficiency by up to 30%. While pricier, they're ideal for larger installations and varying weather conditions.

Harmony in the System

Inverters and controllers may not have the shine of solar panels, but their significance is undeniable. They're like the rhythm section in a band – not always in the spotlight, but essential to the harmony of the music. In this case, the music is the seamless, efficient conversion and flow of solar energy, powering our world with the sun's abundant rays.

As we integrate solar energy more into our lives, appreciating and understanding these components becomes paramount. They remind us that every part, no matter how hidden or inconspicuous, plays a crucial role in the symphony of sustainable energy.

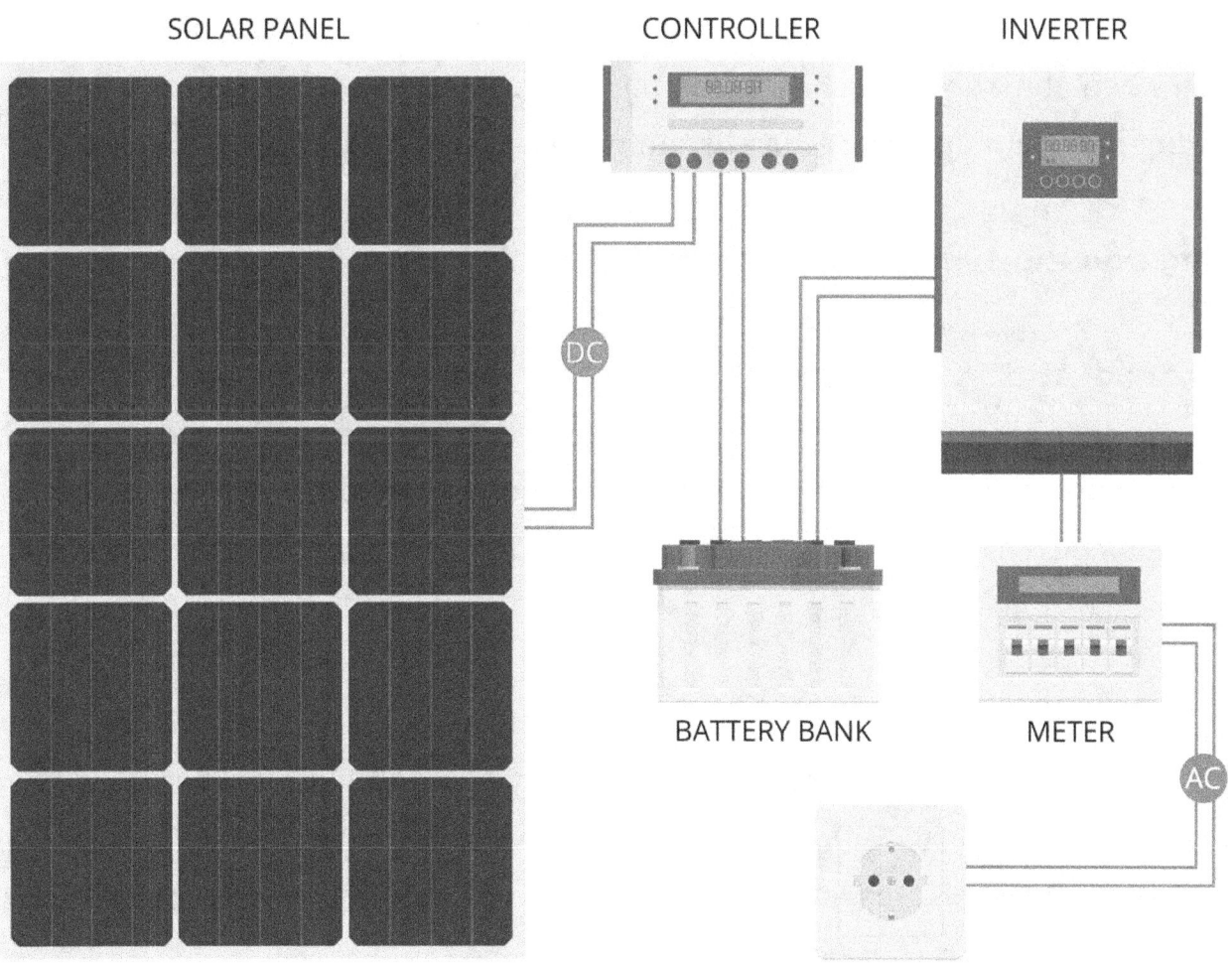

Batteries and Storage Solutions

As the day wanes and the vibrant hues of twilight drape the sky, a home powered by solar energy doesn't retreat into darkness. The reason? Batteries. They silently store the sunlight, ensuring that the energy harnessed during the day continues to light up homes, power devices, and heat water long after the sun sets.

Understanding Battery Basics

Batteries for solar systems are not your regular AA cells. These are robust, specialized devices designed to store significant amounts of energy and discharge it smoothly over time.

The basic principle is straightforward: when your solar panels produce more electricity than your home needs, the excess energy is used to charge the batteries. Then, when the sun isn't shining, or during peak usage times when your panels can't keep up with demand, your system draws energy from these batteries.

Diving into the Types of Solar Batteries

1. **Lead-Acid:** A tried-and-true technology, lead-acid batteries have been used for over a century. While they are relatively inexpensive and widely available, they tend to be bulky and have a shorter lifespan compared to newer technologies. They're often used in off-grid setups where their size and weight aren't significant issues.

2. **Lithium-Ion:** The same technology that powers your smartphone or laptop can also be used for your home's solar system. Lithium-ion batteries have a higher energy density, meaning they can store more energy in a smaller space. They also have a longer lifespan and better efficiency than lead-acid batteries, but they tend to be more expensive.

3. **Flow Batteries:** A newer entrant, flow batteries store energy in a liquid form. They have a unique advantage: their energy capacity (how much energy they can store) is independent of their power rating (how quickly they can release energy). This means they can be easily scaled up for industrial applications. However, their complexity can make them more expensive and less suited for residential use.

4. **Saltwater Batteries:** These batteries use saltwater electrolytes to store energy. They're relatively new and offer the benefits of being non-toxic and having a long cycle life. However, they're still in the early stages of commercialization and might be less readily available.

Considering Battery Lifespan and Depth of Discharge (DoD)

Every battery has a specified lifespan, usually denoted in cycles. A cycle consists of a battery being charged up and then discharged. For instance, a battery with a lifespan of 5,000 cycles can be charged and discharged 5,000 times before its capacity significantly diminishes.

Another crucial term is the Depth of Discharge (DoD). It indicates the percentage of a battery's energy that has been discharged relative to the overall capacity. A higher DoD usually means you can use more of your battery's capacity without affecting its lifespan adversely.

For example, if a 10 kWh battery has a DoD of 90%, you can use 9 kWh of energy without harming the battery's longevity. It's essential to balance the DoD with your energy needs to get the most out of your storage solution.

Incorporating Storage for a Sustainable Tomorrow

In the context of a world grappling with climate change and seeking sustainable alternatives, the importance of efficient storage cannot be understated. Batteries bridge the gap between the sun's intermittent availability and our constant energy needs. They're like the reservoirs in a dam, holding onto the water (or in this case, sunlight) and releasing it steadily, ensuring a constant flow.

CHAPTER 3: Understanding Solar Energy Metrics

Measuring Solar Irradiance

Imagine stepping out on a sunny day, feeling the warmth on your face. It's more than just a comforting sensation; it's a manifestation of energy—solar energy. But how do we quantify this energy, ensuring that our solar systems are set up optimally to harness the maximum power from our radiant sun?

Enter solar irradiance, a term that sounds technical but boils down to a simple concept: measuring the power of the sun.

Unveiling Solar Irradiance

At its core, solar irradiance is the power received from the sun at a particular area over a specific duration. It's usually measured in watts per square meter (W/m^2). Think of it as the sun's intensity. Knowing the irradiance value is crucial for designing and optimizing solar systems. It helps in determining how many solar panels one would need and how efficiently those panels might work throughout the day.

Global, Direct, and Diffuse Irradiance

Solar irradiance isn't just a single monolithic number; it breaks down into different components, each vital for understanding solar power's nuances:

1. **Global Irradiance (GHI):** This is the total sunlight, both direct and scattered, that hits a surface horizontal to the ground. It's a combination of the next two types of irradiance.

2. **Direct Normal Irradiance (DNI):** Picture the sunlight that comes straight down, not deviating or scattering, but hitting a surface perpendicular to the sun. That's DNI. It's the most intense form of solar irradiance and is particularly relevant for concentrated solar power systems.

3. **Diffuse Horizontal Irradiance (DHI):** When sunlight hits the atmosphere, it doesn't just travel straight. It scatters, reflecting off particles, clouds, and even air molecules. The sunlight that reaches us without directly pointing to the sun is the DHI. For regions with overcast conditions, DHI plays a crucial role in determining solar feasibility.

Solar Noon and Peak Sun Hours

"Solar Noon" doesn't necessarily align with 12:00 PM on our clocks. It refers to the time of day when the sun is at its highest point in the sky for a specific location. Why is this important? It's usually when DNI is at its peak, and knowing this time helps in optimizing the orientation of solar installations.

Further, while we talk in terms of 24-hour days, for solar energy, we often refer to "peak sun hours." It's an average number representing the equivalent hours per day when solar irradiance averages 1,000 W/m^2. For instance, if a location has 5 peak sun hours, it means that the energy received during these hours is equivalent to five hours of continuous irradiance at 1,000 W/m^2.

Tools of the Trade: Pyranometers and Solarimeters

To measure these various forms of irradiance, specialized instruments are used:

- **Pyranometers:** These devices measure global irradiance (GHI). They have a hemispherical field of view and can gauge both direct and diffuse sunlight. Calibration and maintenance are essential for these instruments to ensure accurate readings.

- **Solarimeters:** More of a general term, these are tools that measure solar radiation. Depending on the type, they might measure GHI, DNI, or even DHI.

Deciphering Nature's Clues

Solar irradiance isn't static. It varies throughout the day, with the seasons, and even due to atmospheric conditions. By understanding and measuring it, we become better equipped to harness the sun's potential fully. In a world leaning into renewables, comprehending these intricacies of sunlight is not just for the academically inclined; it's for everyone who dreams of a brighter, greener future.

Calculating Energy Needs

The intricate dance of light that we observe daily, with the sun taking its slow ascent and descent, influences not just the circadian rhythms of nature, but the very foundation of solar power systems. But for one to harness this celestial energy, we must first comprehend our energy needs. How do we navigate this maze? How do we quantify our reliance on electrical devices and transform it into data that paves the way for efficient solar setups?

Beginning with the Basics: Your Daily Energy Consumption

It all starts at home—quite literally. Every electrical device, be it your refrigerator humming silently, the laptop that's been your work buddy, or the comforting warmth of your heater, has an energy rating. It's usually denoted in watts (W) or kilowatts (kW). This rating tells you the power it consumes when operational.

To determine the daily energy consumption of a device, multiply its power rating (in kW) by the hours it runs daily. The result gives you the energy used, expressed in kilowatt-hours (kWh).

For instance, a 100W bulb running for 5 hours consumes: 0.1 kW x 5 hours = 0.5 kWh.

Do this for all your appliances, and you get a clearer picture of your daily energy needs.

The Cumulative Effect: Monthly and Yearly Energy Consumption

While daily calculations give you a snapshot, understanding monthly and yearly consumption aids in designing a solar system tailored to your needs. By aggregating daily consumption, you can identify patterns, noting seasons of high or low energy use, and anticipate your solar system's performance throughout the year.

Factoring in Energy Losses

Energy is a tricky customer. Even after we've calculated our needs and designed a system to match, losses occur. These can be due to inefficiencies in the solar panels, energy losses in the inverter, or even transmission losses as electricity travels through wires. Typically, one might factor in a loss of around 10-20%, but this can vary based on the quality of your equipment and installation.

Peak Demand: Preparing for the Energy Spikes

There will be times when your energy consumption spikes. It could be during a particularly cold day when heaters run incessantly, or when hosting a gathering, with every light and sound system in operation. These peaks in energy demand are crucial to note. A well-designed solar system should handle not just average consumption but these occasional spikes without faltering.

Buffer for the Rainy Days: Oversizing for Assurance

Relying entirely on solar power necessitates a system that delivers, even on days when the sun is shy. By "oversizing" a solar system slightly (installing more capacity than average needs dictate), you provide a buffer. It ensures that even during cloudy days or unexpected high-consumption scenarios, your solar setup won't leave you in the dark.

The Roadmap to Energy Independence

As we unravel the threads of our energy consumption, layer by layer, we're not just indulging in a mathematical exercise. We're charting a course towards energy independence. Calculating energy needs isn't just about numbers; it's about understanding our relationship with electricity, recognizing our patterns, and crafting a solar solution that mirrors our lifestyles.

CHAPTER 4: Positioning and Tracking

Solar Panel Angles and Efficiency

When we talk about solar energy, it's not merely about panels and wires. It's about a dance—a choreography between the sun and the structures designed to harness its power. The sun, while powerful, is also elusive. Its trajectory changes, influenced by the seasons and our planet's tilt. Hence, the angle at which solar panels are positioned isn't a trivial detail; it's the cornerstone of efficient energy capture.

The Magic of the Optimal Angle

The sun's path varies based on two primary factors: the latitude of your location and the time of year. The "optimal angle" is the inclination at which a solar panel should be tilted to capture the maximum sunlight during the day.

For those living in the equatorial regions, where the sun is almost directly overhead, a lesser tilt might be ideal. As we move towards the poles, the angle of the sun's rays becomes more slanted, demanding a steeper panel tilt.

But why does this matter? A solar panel's efficiency hinges on how perpendicularly sunlight hits it. The closer the sunlight is to a 90-degree angle, the more energy the panel can generate. This relationship between light incidence and energy production underscores the importance of finding that "sweet spot" of tilt.

Seasonal Adjustments: Adapting to the Sun's Journey

As Earth orbits the sun, the sun's perceived position in our sky shifts. This celestial movement means that a fixed solar panel angle might not be the most efficient year-round. Adjusting the tilt of the panels according to the seasons can lead to a significant boost in energy capture.

For example, during winter months in the Northern Hemisphere, the sun is lower in the sky. A steeper tilt during these months ensures that panels catch more direct sunlight. Conversely, in summer, when the sun is higher, reducing the tilt can optimize energy production.

Fixed vs. Adjustable Mountings

Solar installations can broadly be categorized into two based on their mountings:

1. **Fixed Mountings**: These setups have solar panels mounted at a constant angle. While less flexible, they are also less expensive and require minimal maintenance.

2. **Adjustable Mountings**: These allow the angle of the panels to be changed periodically. Though they might demand more attention and come with a slightly higher price tag, the potential for increased energy capture can make them a worthy investment, especially in regions with pronounced seasonal variations.

The Balance of Aesthetics and Efficiency

While the angle is paramount for efficiency, it's not the only consideration. For many homeowners, the aesthetics of solar installations matter. Sleek, flush-mounted panels that align with the roof's slope might be visually pleasing, but they may not always be at the optimal angle for sunlight capture. It's a delicate balance, and the right choice often hinges on individual priorities and the architectural characteristics of the building.

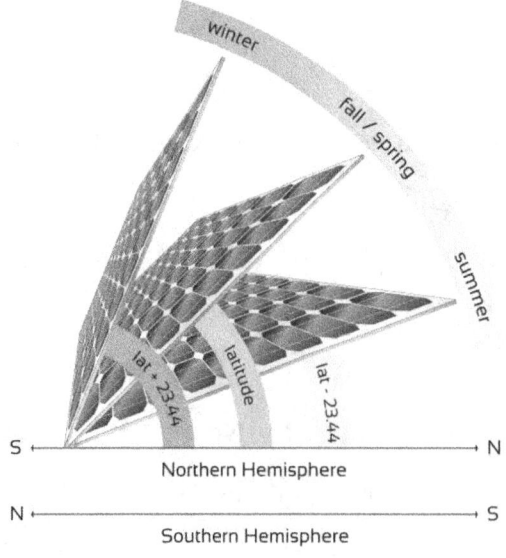

The Role of Solar Trackers

If positioning the solar panels at the right angle is akin to setting the sails of a ship to catch the wind, then adding solar trackers is like having a crew that adjusts the sails as the wind changes direction. While the wind's course might be unpredictable, the sun's journey across the sky is consistent. This predictability is what solar trackers exploit to ensure panels are always facing the sun head-on, maximizing their efficiency.

The Mechanics Behind Solar Trackers

At their core, solar trackers are motor-driven mounts equipped with sensors. These sensors detect the sun's position and send signals to the motors to adjust the panel's orientation. Think of them as the "navigators" for the solar panel "ship," always ensuring it's on the optimal course.

There are primarily two types of trackers:

1. **Single-Axis Trackers**: These move panels either from east to west or follow the sun's daily path. They're most effective in regions closer to the equator where the sun's movement is predominantly horizontal.

2. **Dual-Axis Trackers**: More advanced and able to move panels both horizontally and vertically, these trackers follow the sun's daily east-to-west trajectory and account for its changing height in the sky through different seasons.

Boosting Solar Output: Efficiency Meets Precision

By continuously adjusting the panel's orientation, solar trackers can substantially increase a solar system's energy output. Compared to fixed panels, systems equipped with trackers can experience efficiency boosts of up to 25% for single-axis trackers and as much as 40% for dual-axis trackers.

It's like giving the solar panels "eyes" to watch the sun and "limbs" to turn towards it, ensuring every possible photon is captured and converted.

The Complexities of Solar Tracking

While trackers bring the promise of higher efficiency, they also introduce mechanical complexity to solar installations:

- **Maintenance**: Since trackers have moving parts, they might require more frequent checks and maintenance compared to fixed mounts.

- **Cost**: Trackers, especially dual-axis ones, can significantly increase the initial investment needed for a solar setup.

- **Space**: Given that panels move throughout the day, solar installations with trackers demand more spacing between panels to prevent shading.

To Track or Not to Track?

While the promise of boosted efficiency is tempting, solar trackers aren't a universal solution. For regions near the equator, where the sun's path is more horizontal, single-axis trackers might offer a good return on investment. However, in regions with pronounced seasonal changes, dual-axis trackers can be more beneficial.

Yet, it's essential to weigh the benefits against the added costs and complexities. For some setups, especially smaller residential ones, the increased output might not justify the added investment in trackers.

A Dance with the Sun

With or without trackers, the core philosophy remains unchanged: harnessing the sun's immense power for a sustainable future. Trackers are merely tools in this grand endeavor, amplifying what's already an extraordinary feat of human ingenuity.

As we envision a world powered by the sun, we must consider every tool, every technique, and every innovation that brings us closer to that radiant dream.

BOOK 3

Planning and Designing Your Solar System

CHAPTER 1: Site Analysis

Identifying Solar Potential

Before an artist begins to paint, they study the canvas. The same principle applies to installing a solar system. The landscape or structure where you plan to set up the system is the canvas. By understanding its nuances, strengths, and limitations, you can optimize your solar installation for maximum efficiency.

What is Solar Potential?

Solar potential refers to the capability of a specific location or site to harness sunlight. Not every spot on your property will have the same solar potential. Buildings, trees, or other obstructions can cast shadows, seasonal changes can alter sun angles, and regional weather patterns can affect solar exposure.

Key Factors in Identifying Solar Potential

1. **Solar Insolation**: This is the amount of solar radiation received at a particular location. Measured in kWh/m^2/day, it gives a clear idea of how much sunlight a site gets. High insolation indicates more potential for solar energy generation.

2. **Orientation**: The direction your panels face will significantly impact their energy output. In the Northern Hemisphere, south-facing installations generally receive the most sunlight.

3. **Tilt Angle**: This refers to the angle at which panels are positioned relative to the ground. Ideally, this angle should be adjusted seasonally to match the sun's height in the sky.

4. **Shading**: Even brief periods of shade can significantly reduce a solar panel's performance. It's crucial to ensure that the chosen site is free from obstructions like trees, buildings, or chimneys that might cast shadows.

Assessment Tools and Techniques

Solar Maps: Many regions now have solar maps, which can quickly give you a general idea of the solar potential in your area. These maps take into account historical weather data and average insolation values.

Site Surveys: A thorough site survey involves an expert visiting your property to assess its solar potential. They'll check for shading obstructions, measure available space, and note the orientation and tilt options.

Solar Pathfinder: This tool helps in understanding how shadows will move across a site throughout the year. It combines the site's specific shade patterns with solar insolation data to give a comprehensive solar potential analysis.

How Site Characteristics Influence System Design

Each site's unique characteristics will guide the solar system design. For instance, if a roof has multiple levels or angles, it might be beneficial to use micro-inverters, which allow individual panels to operate independently, ensuring that one shaded panel doesn't drag down the performance of others.

Ground-mounted systems, on the other hand, offer more flexibility in orientation and tilt but might require landscaping or terracing to optimize the site.

Embracing Nature's Variability

While it's essential to maximize a site's solar potential, it's equally crucial to remember that nature is variable. Days will come when clouds dominate or snow blankets the panels. What's important is the average sunlight received over time, ensuring consistent energy generation and savings.

The Foundation of a Successful Solar Journey

Identifying the solar potential is the foundational step in the solar installation process. It informs every subsequent decision, from system size to component selection. By understanding the unique tapestry of sunlight, shade, and space at a given site, one can weave together a solar system that not only generates energy but resonates with the rhythms of nature.

Shadow Analysis

It's a dance as old as time itself: the interplay between light and shadow. When dealing with solar installations, shadows become an especially significant consideration. They can play a significant role in how much energy a solar system can produce, making shadow analysis an essential aspect of solar site planning.

Why Shadow Analysis is Crucial

A solar panel needs sunlight to generate electricity. Any object that obstructs the sunlight—whether it's a tree, building, or even a pole—can affect the panel's performance. Even a short duration of shading during peak solar hours can lead to a considerable reduction in energy output.

Understanding the Types of Shadows

1. **Fixed Shadows**: These are caused by permanent structures, such as buildings or tall walls. Their position doesn't change with time, but the shadow they cast can vary with the sun's angle.

2. **Dynamic Shadows**: Objects like trees, which can grow over time, or structures like flags that move with the wind, cause dynamic shadows. These are unpredictable and can vary in their impact on a solar setup.

Techniques and Tools for Shadow Analysis

Solar Pathfinder: As previously mentioned, a Solar Pathfinder provides insights into the sun's path across your site. It superimposes the site's potential shading onto a sunpath diagram, giving a month-by-month breakdown of sun exposure.

Digital Simulation: Modern software allows for digital simulations of a site's shadow patterns. By inputting the dimensions and locations of potential shading objects, these tools provide a detailed year-round analysis.

Drone Surveys: Equipped with cameras, drones can capture high-resolution images of a site from various angles. When processed, these images can offer valuable shading insights, especially for larger properties or installations.

Making Adjustments Based on Shadow Analysis

Once you've identified potential shading objects and their impact, several adjustments can be made:

1. **Panel Placement**: If shadows affect only a specific area, consider shifting your solar panel layout to a sunnier spot.

2. **Tree Trimming or Removal**: Sometimes, strategic trimming of tree branches can mitigate shading. In extreme cases, it might be necessary to consider removing a tree. However, this should be a last resort due to the ecological benefits trees provide.

3. **Opt for Micro-Inverters**: Unlike traditional string inverters, micro-inverters operate on a per-panel basis. This means that even if one panel is shaded, the performance of others remains unaffected.

4. **Elevated Mounting**: By raising the height of solar panels, some shading issues can be minimized. This is especially useful for ground-mounted solar installations.

The Art of Compromise

Sometimes, it might not be possible to eliminate all shadows. In such cases, it's about striking a balance. Calculate the energy loss due to shading and weigh it against the cost and feasibility of making adjustments. It might be more practical to accept a slight reduction in efficiency rather than undertake significant site modifications.

Shadows and Seasons

The angle and intensity of the sun's rays change with the seasons, and so do shadows. A site that's sunny during the summer might experience more shading in the winter when the sun is lower in the sky. Seasonal shadow analysis ensures the solar system performs optimally throughout the year.

CHAPTER 2: System Sizing and Scalability

Calculating System Size

Venturing into the world of solar energy is not just an eco-friendly choice; it's also a wise economic decision. However, to truly harness the power of the sun, it's essential to install a solar system tailored to your specific needs. The size of your solar system is, arguably, the most critical aspect of this endeavor.

Determining Your Energy Consumption

Before delving into the nitty-gritty of sizing your solar system, understanding your energy consumption is paramount. It's the foundation upon which the rest of the calculations stand.

- **Review Your Utility Bills**: A close examination of your utility bills from the past year provides a solid estimation of your monthly energy consumption. This gives a comprehensive picture, accounting for seasonal fluctuations in energy use.

- **List Major Appliances and Their Consumption**: By listing all significant power-consuming appliances in your property and calculating their daily energy use, you can estimate your daily consumption.

Evaluating Solar Potential

Every location has its own solar potential, influenced by its latitude, average number of sunny days, and other climatic factors.

- **Sun Hours**: Not every hour of daylight is optimal for solar energy production. Sun hours refer to the number of hours when solar irradiance exceeds a certain level, typically considered as the hours when the sun can effectively produce solar energy.

- **Solar Insolation**: This is the amount of solar radiation received per unit area. It varies by location and helps in determining the energy your solar panels can produce.

Calculating the System Size

With a grasp on your energy needs and the solar potential of your location, you can now calculate the required system size.

1. **Determine Daily Energy Needs**: Sum up your daily energy consumption. For instance, if your monthly consumption is 300 kWh, your daily consumption would be roughly 10 kWh (300 kWh/30 days).

2. **Factor in the Sun Hours**: If your location gets an average of 5 sun hours per day, and you need to produce 10 kWh daily, you'd need a system that can produce 2 kW per hour (10 kWh/5 hours).

3. **Account for Efficiency Losses**: No system is 100% efficient. Depending on the components and local conditions, the overall efficiency might be 80-90%. Therefore, if you need a system that produces 2 kW, considering an 85% efficiency, the actual requirement would be around 2.35 kW.

The Importance of Scalability

While understanding your current energy needs is vital, it's equally crucial to consider future requirements.

- **Anticipate Growth**: If you're planning on expanding your property, adding more electrical appliances, or if your family is growing, your energy requirements will increase.

- **Plan for Future Technology**: As the world leans more towards electrification, especially with the rise of electric vehicles, your future self might thank you for having a solar system that can accommodate these technologies.

Tools and Professional Assistance

While these steps provide a basic understanding of system sizing, it's always beneficial to employ tools or seek professionals for a detailed analysis.

- **Solar Calculators**: Numerous online tools can provide a rough estimate of the required solar system size based on location and energy consumption.

- **Professional Solar Auditors**: These experts can conduct a thorough onsite analysis, considering all variables, and provide a comprehensive solar solution tailored to your needs.

Future-proofing Your System

Future-proofing refers to the anticipation of future developments and ensuring that your current choices remain relevant and effective in the coming years. In the context of solar energy systems, it's about designing a system that can efficiently adapt to potential changes in energy consumption and technology advancements.

Why Future-proof Your Solar System?

- **Economic Sense**: While upgrading a system down the line is possible, integrating flexibility from the outset is often more cost-effective. Retrofitting or expanding an existing system can come with logistical challenges and added expenses.

- **Environmental Stewardship**: A future-proof system is inherently more sustainable. By reducing the need for future manufacturing and installation processes, you're effectively cutting down on the associated environmental footprint.

- **Peace of Mind**: Knowing that your system is prepared for future scenarios provides a sense of security, eliminating the worry of potential energy shortfalls or cumbersome upgrade processes.

Strategies to Future-proof Your Solar Installation

1. **Oversizing Inverters**: While it might sound counterintuitive, installing an inverter that's rated higher than your current panel capacity can be a smart move. This provides the flexibility to add more panels in the future without needing a new inverter.

2. **Space Considerations**: If space permits, it's wise to leave room for additional panels when setting up your initial installation. Even if you don't plan on expanding the system soon, having the option can prove invaluable later.

3. **Opt for Modular Systems**: Some solar setups are designed with modularity in mind, allowing for easy expansion. These systems can be scaled up or down based on changing needs, making them ideal for future-proofing.

4. **Stay Updated with Technology**: The solar industry is continuously evolving, with new technologies emerging frequently. By staying informed, you can take advantage of breakthroughs that can be integrated into your system.

5. **Robust Infrastructure**: Invest in durable mounting systems and high-quality wiring that can accommodate potential expansions. This ensures that the foundational elements of your solar setup are ready for the future.

6. **Energy Storage**: As battery technologies improve and become more affordable, integrating energy storage solutions can add another layer of future-proofing. Not only do they provide backup power, but they also offer flexibility in terms of energy usage and potential grid interactions.

7. **Monitoring and Maintenance**: Regularly monitor your system's performance. Many modern systems come with sophisticated monitoring tools that provide insights into energy production and consumption. By keeping an eye on these metrics, you can preemptively address issues and adapt to changing energy patterns.

The Role of Policy and Incentives

While individual choices play a significant role, it's also essential to be aware of broader policy changes and incentives. Governments worldwide are recognizing the importance of renewable energy and are offering incentives for expansions or upgrades. Staying informed about these can provide both financial and logistical support in future-proofing your system.

CHAPTER 3: Roof vs. Ground Mounted Systems

Pros and Cons of Roof Mounting

As urban landscapes become ever more congested and ground space turns into a premium commodity, the rooftops of our homes and buildings offer a tantalizing promise: vast, often underutilized spaces that gaze directly up at the sun. Roof-mounted solar panels, consequently, have become the go-to for many urban dwellers and businesses. But as with all choices, this method comes with its set of advantages and challenges.

The Lure of Altitude: Advantages of Roof Mounting

- **Space Efficiency**: In crowded urban settings, using the roof for solar installations is a way to utilize space that might otherwise go to waste. Instead of taking up precious yard or ground space, the panels perch atop structures that already exist.

- **Cost-Effectiveness**: Often, roof-mounted systems might prove to be less expensive than ground-mounted ones. This is because they don't necessitate additional structures or extensive groundwork, relying instead on the existing building's framework.

- **Aesthetics**: For those concerned about the visual impact of solar panels, roof mounting can offer a more discreet option. Integrated solar roofs, for instance, blend seamlessly with the architecture, ensuring the technology doesn't overshadow the aesthetic appeal of a building.

- **Safety and Security**: Elevated from the reach of casual passersby or potential vandals, roof-mounted panels are less prone to tampering or theft.

- **Potential Cooling Benefits**: The shade provided by the panels can lead to a cooler roof surface, which in turn can help reduce the heat entering the building, especially in hot climates.

Scaling Heights: The Challenges of Roof Mounting

- **Roof Condition and Longevity**: Not all roofs are created equal. Before considering a roof-mounted solar system, the structure's age, material, and condition must be assessed. An older roof might require repairs or replacement before it can bear the weight of solar panels.

- **Orientation and Tilt**: While ground-mounted systems can be easily oriented for optimal sun exposure, roof-mounted panels are at the mercy of the building's design. If the roof doesn't face the ideal direction or has an unsuitable tilt, the efficiency of the panels may be compromised.

- **Maintenance and Cleaning**: Elevated off the ground, roof panels can be harder to access for cleaning, maintenance, or repairs. This might lead to increased maintenance costs or a reliance on professionals for routine tasks.

- **Potential Roof Damage**: While rare, there's always the risk of the roof getting damaged during the installation process. Proper sealing and waterproofing are crucial to prevent leaks or structural issues later on.

- **Heat Accumulation**: In some cases, especially without proper ventilation, the space beneath roof-mounted panels can become a heat trap, potentially leading to increased indoor temperatures.

Customizing Roof Mounts: Variations to Consider

Solar technology has evolved to offer solutions tailored to specific roof types and challenges:

1. **Flush Mounts**: Ideal for pitched roofs, these are the most common type of mounts, allowing panels to sit at a fixed angle.

2. **Tilt Mounts**: For flat roofs or those with a less-than-ideal pitch, these mounts allow the panels to be tilted for better sun exposure.

3. **Ballasted Footing Mounts**: These are non-penetrative mounts that rely on the weight (or ballast) to hold the panels in place, reducing the risk of roof damage.

4. **Integrated Solar Roofs**: These are solar panels designed to function as the roof itself, replacing traditional roofing materials. They're both functional and aesthetically appealing.

Making the Choice: Is Roof Mounting for You?

While the idea of utilizing roof space for solar energy is compelling, it's vital to conduct a comprehensive assessment of your specific situation. Consider the age, condition, and orientation of your roof. Factor in future maintenance needs, potential costs, and the local climate.

Roof-mounted solar systems offer a blend of efficiency, aesthetics, and practicality. However, they demand careful planning and consideration to truly shine. As we explore further into ground-mounted systems, a clearer picture will emerge, helping homeowners and businesses make an informed decision in their solar journey.

Benefits of Ground Installations

The vast horizons stretching beyond our homes and buildings often beckon with an enticing proposition: what if the ground, from which life springs, becomes the bedrock for harnessing the sun's power? Ground-mounted solar systems stand as a testament to this thought, providing an alternative to the lofty reaches of rooftops. They come with their own set of merits and considerations, offering unique advantages in the quest for sustainable energy.

Sun-Kissed Earth: Advantages of Ground Installations

- **Optimal Orientation**: Unlike the fixed orientation of a roof, ground-mounted systems offer greater flexibility. They can be placed directly facing the sun, ensuring optimal energy absorption. Adjustable tilt mechanisms can further fine-tune the panels' angle to seasonal changes.

- **Scalability**: Ground installations can easily be expanded. Whether you have an increasing energy need or wish to add more panels later, the ground offers a limitless canvas for scalability.

- **Efficient Cooling**: Ground-mounted panels tend to stay cooler than roof-mounted ones due to better airflow around them. Cooler panels are generally more efficient and have a longer lifespan.

- **Ease of Maintenance**: Ground-level accessibility means easier cleaning, maintenance, and monitoring. There's no need to climb a ladder or navigate a rooftop to tend to your panels.

- **No Structural Concerns**: Ground installations sidestep the challenges of assessing and potentially reinforcing existing structures. They stand independent of your home or building, eliminating concerns about roof age or material.

- **Dual-Purpose Land Use**: In a practice called agrivoltaics, agricultural activities and solar power generation can co-exist. Crops, livestock, and solar panels can share space, allowing for productive land use.

Ground Realities: The Challenges of Ground Mounting

- **Land Use**: Ground installations demand space. For those with limited land, this can be a prohibitive factor. It's essential to ensure that the solar setup doesn't disrupt other land uses or natural habitats.

- **Potentially Higher Initial Costs**: While they might save costs in the long run, the initial investment for ground-mounted systems, including the construction of mounting structures and potential land grading, can be higher than roof installations.

- **Permitting and Zoning**: Ground-mounted installations might face stricter regulations, zoning laws, or permit requirements, especially in densely populated or environmentally sensitive areas.

- **Security Concerns**: Being at ground level, these systems can be more accessible to potential vandals or wildlife. Proper fencing or security measures might be necessary to protect the investment.

Choosing Your Grounded Setup: Types and Variations

Solar technology isn't a one-size-fits-all solution. Ground-mounted systems come in various configurations, catering to diverse needs:

1. **Standard Ground Mounts**: These are driven into the ground and hold up panels at a fixed angle. They are typically made of aluminum and can be adjusted to the desired angle.

2. **Pole Mounts**: These elevate panels higher off the ground using a singular pole. They can support multiple solar panels and sometimes come with a tracking system to follow the sun's movement.

3. **Solar Trackers**: While more expensive, trackers allow panels to move with the sun throughout the day, maximizing energy absorption. They can increase a solar system's efficiency by 25% or more.

Weighing the Decision: Ground Installations in Focus

Ground-mounted solar installations are a testament to human adaptability, turning patches of earth into powerhouses of energy generation. They're particularly suitable for those who have land to spare, or for commercial entities with expansive areas.

Before embarking on this route, it's essential to conduct a thorough land assessment, understand local regulations, and gauge the practicality of the installation. As with all solar solutions, the aim is to harmonize with the environment, ensuring that the embrace of technology complements the rhythms of nature.

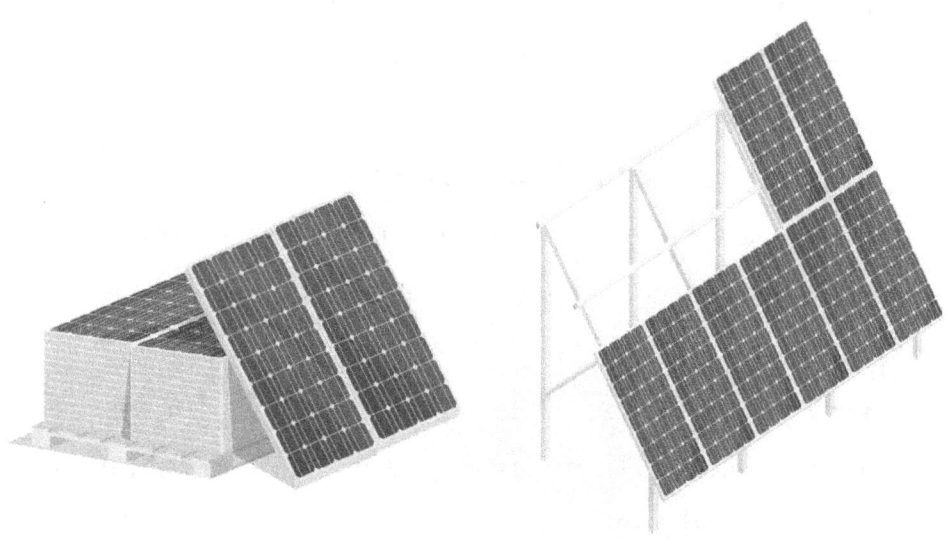

CHAPTER 4: Wiring and Safety Protocols

Solar System Wiring Basics

Solar installations might seem, at first glance, a mere assembly of panels soaking up the sun's rays. But behind the sunlit facade is a complex web of wires, connectors, and conduits, ensuring that the energy captured is safely and efficiently routed to power our homes and appliances. This network of wires, much like the human nervous system, is crucial for the optimal functioning of the solar system. Understanding the basics of this wiring can be the key to unlocking the full potential of your solar setup.

From Sun to Socket: The Flow of Energy

When photons from sunlight strike the solar panels, a DC (direct current) voltage is generated. This DC power needs to travel through a series of components before it's transformed into usable AC (alternating current) power for most homes. This journey is facilitated by the system's wiring.

1. **From Panel to Inverter**: After being generated in the solar panels, the DC power is carried through a set of wires to the inverter. These wires are typically encased in a protective sheath to prevent exposure and damage. They are also color-coded (positive usually in red and negative in black) to ensure proper connections.

2. **The Role of the Inverter**: The inverter's job is to convert the DC power from the panels into AC power. It's connected to both the solar panels and the home's electrical system.

3. **Distribution Panel and Grid**: Post-inversion, the AC power is sent to the home's main distribution panel, from which it's distributed to different parts of the house. Any excess energy is often sent back to the grid, especially in grid-tied systems.

Key Components in Solar Wiring

- **Solar Cables**: These are specially designed cables that can withstand the outdoor environment of a solar installation. UV resistant and waterproof, these cables connect solar panels to the inverter.

- **Connectors**: To prevent mishaps and ensure smooth energy flow, connectors play a crucial role. MC4 connectors are a common standard in the industry, known for their secure and waterproof connection.

- **Conduits**: These are protective tubes that house the wiring, offering protection against environmental factors and potential damage.

- **Junction Boxes**: Often placed behind solar panels, these boxes are where the panel's wiring terminations are made and provide an easy point of access for maintenance.

Safety First: Considerations in Solar Wiring

- **Voltage Handling**: Ensure that the wiring and components can handle the system's voltage. Using wires and connectors rated for higher voltages offers a margin of safety.

- **Weather Resistance**: Given the outdoor nature of solar installations, it's essential to use wiring components resistant to UV rays, rain, and other environmental factors.

- **Grounding**: Proper grounding is vital for safety. It prevents electric shock hazards and potential damage to equipment.

- **Color Coding and Labeling**: Always adhere to color-coding standards. Proper labeling can aid in maintenance and troubleshooting.

- **Regular Inspection**: Periodically inspect the wiring for any signs of wear, tear, or damage. Ensure that the conduits and protective sheaths are intact.

Safety Measures and Precautions

Harnessing the power of the sun to electrify our homes and businesses is undeniably revolutionary. Yet, like all electrical systems, solar installations come with their own set of hazards and challenges. The emphasis on safety cannot be overstated. When we approach solar with caution, respect, and knowledge, we can ensure not only the system's longevity but also the well-being of those who interact with it.

Potential Hazards: Recognizing the Risks

Before diving into the precautions, let's briefly outline the potential hazards associated with solar installations:

- **Electrical Shocks**: As with any electrical system, there's a risk of shock. Incorrect wiring, damaged components, or touching live wires can result in electric shocks, which can be fatal in high-voltage systems.

- **Fire Hazards**: Faulty connections, damaged wires, or a compromised inverter can pose a fire risk.

- **Chemical Hazards**: Batteries, especially older lead-acid ones, can leak or explode if mishandled, releasing harmful chemicals.

- **Physical Injuries**: Installing panels, especially on rooftops, can lead to falls or injuries from dropped tools or materials.

Preventive Measures: Minimizing Risks

- **Professional Installation**: Always consider hiring a certified professional for installation. Their expertise can help in ensuring that the system is safely and effectively set up.

- **Regular Maintenance**: Scheduled inspections and maintenance can help identify and rectify potential issues before they escalate. Look for corroded parts, damaged wires, or any irregularities in the system.

- **Safety Gear**: Wear appropriate safety gear during installation or maintenance. This includes gloves to protect against electric shocks, safety harnesses for rooftop work, safety goggles, and protective footwear.

- **Disconnect Before Maintenance**: Before embarking on any repair or maintenance, ensure the system is disconnected from the grid and other power sources.

- **Battery Care**: If your system includes batteries, ensure they are in a well-ventilated area. Regularly check for signs of damage or leakage. Handle batteries with care, using proper safety gear.

- **Clear Signage**: Clearly label all components, especially switches and breakers. This can aid in quick identification during emergencies or routine checks.

Safety Protocols: Ensuring Systematic Safety

- **Emergency Shutdown Procedure**: Have a clear procedure for shutting down the system in emergencies. Ensure all members of the household or facility are familiar with it.

- **Training**: Offer training sessions for those living or working around the solar installation. Awareness can be a significant factor in preventing accidents.

- **Keeping the Area Clear**: Ensure the area around solar components, especially inverters and batteries, is free from flammable materials. This reduces the risk of fire.

- **Documentation**: Keep a record of all installations, maintenance checks, and repairs. This log can be invaluable for troubleshooting and ensuring regular safety checks.

Staying Updated: The Changing Landscape of Safety

Solar technology is ever-evolving, and so are the safety measures associated with it. Stay updated with the latest safety guidelines and technological advancements. Subscribing to industry journals, attending workshops, or being part of a solar community can provide invaluable insights.

BOOK 4

DIY Solar Installation Guide

CHAPTER 1: Preparing for Installation

Necessary Tools and Equipment

While the ambition and drive to incorporate sustainable energy sources into your life are commendable, the foundation of a successful installation lies in meticulous preparation. And the crux of that preparation? Equipping yourself with the right tools and understanding their purposes.

The Toolset: From Basics to Specifics

Every artisan requires their set of tools, and for those embarking on a solar installation journey, the toolkit is both varied and vital.

- **Multimeter**: An essential diagnostic tool, the multimeter helps in measuring voltage, current, and resistance. With solar installations, ensuring the correct voltage and current from panels is paramount, and a multimeter assists in these measurements.

- **Wire Strippers and Crimpers**: Solar installations involve a lot of wiring. Having a reliable set of wire strippers helps ensure clean connections. Crimpers, on the other hand, are crucial for securing terminal rings and connectors.

- **Solar MC4 Connectors and Spanners**: MC4 connectors are standardized connectors for solar panels. They ensure secure connections between panels, which are crucial for the efficient transfer of electricity. The spanners aid in tightening and loosening these connectors.

- **Cable and Conduit Cutters**: Given the lengths of cables and conduits used in solar installations, having dedicated cutters can ease the installation process. Sharp, reliable cutters ensure clean cuts, which in turn lead to better connections.

- **Ladder and Roof Safety Equipment**: If your solar installation is rooftop-based, investing in a sturdy ladder and roof safety equipment, such as harnesses and anchors, becomes essential.

- **Drilling Machine and Bits**: Used for mounting brackets and securing panels, a reliable drilling machine with a variety of bits suited for different materials is invaluable.

- **Labeling Machine**: Given the multitude of connections and wires, labeling becomes crucial. A labeling machine assists in keeping everything organized, making future maintenance and troubleshooting more straightforward.

- **Level and Measuring Tape**: Precision is key in solar installations. Using a level ensures panels are mounted evenly, maximizing their efficiency. A measuring tape helps maintain consistent spacing and accurate placements.

Setting Up Your Workspace

Equally important as the tools themselves is the workspace. A clean, organized area aids in streamlining the installation process.

- **Tool Organizer**: Keeping tools in easily accessible, organized spots can drastically reduce installation time and minimize errors.

- **Portable Workbench**: A dedicated space for laying out tools, making measurements, or pre-assembling components can be a boon.

- **Safety First**: Always ensure your workspace is free of hazards. This includes keeping it free of loose wires, ensuring proper grounding of electrical tools, and using safety gear like gloves and goggles.

Knowledge: The Intangible Tool

While physical tools play a significant role, equipping oneself with knowledge is equally vital. Understanding the components of your solar system, familiarizing yourself with installation guidelines, and being aware of local regulations and permits can save a lot of time and prevent costly mistakes.

Safety Gear and Protocols

The realm of DIY solar installation is filled with excitement and the promise of sustainable energy. Yet, beneath the layers of enthusiasm, there's a foundational aspect that cannot be ignored: safety. As with any electrical installation, the potential for hazards exists. Recognizing and mitigating these risks is essential for a successful and injury-free installation.

Safety Gear: Your First Line of Defense

Solar installation, especially on rooftops, has its unique set of challenges. From potential falls to electrical shocks, the hazards are real, but with the right gear, they can be substantially minimized.

- **Safety Helmets**: Protecting the most vital part of your body, the head, from potential impacts is non-negotiable. A robust helmet can shield you from accidental hits, especially when working in tight or elevated spaces.

- **Electrical Insulation Gloves**: As you'll be dealing with electrical components, it's imperative to have gloves that offer protection against electrical shocks. These gloves are specially designed to insulate and resist electricity, ensuring your hands remain safe.

- **Safety Goggles**: Your eyes are vulnerable to dust, debris, and even sparks. Safety goggles ensure that they remain protected, giving you clear visibility throughout the installation.

- **Roof Harness**: If your solar panels are to be mounted on a rooftop, a roof harness becomes indispensable. It ensures that in the event of a misstep or loss of balance, you remain secure, preventing potential falls.

- **Safety Boots with Grip**: Slip-resistant boots can be a lifesaver, especially when working on inclined rooftops or damp surfaces. They provide the necessary grip, reducing the risk of slips and falls.

- **Ear Protection**: While it might seem superfluous, prolonged exposure to drilling noises can harm your hearing. Earplugs or earmuffs can help in these scenarios.

Protocols: The Blueprint of Safety

Merely possessing safety gear isn't enough. Adhering to safety protocols ensures that the gear serves its purpose, and you remain safe throughout the installation process.

- **Stay Informed**: Before starting, familiarize yourself with the manufacturer's guidelines for each component. They often contain specific safety instructions that are essential.

- **Work in Teams**: Solar installation, especially lifting heavy panels and mounting them, is a task best done with assistance. Beyond the ease it provides, having someone with you ensures that in case of an emergency, help is Immediately available.

- **Isolate and Test**: Before making any electrical connections, ensure that the power sources are isolated. Once connections are made, always test before switching the main power on. This minimizes the risk of electrical shocks.

- **Weather Considerations**: Avoid installing during inclement weather. Wet conditions can increase the risk of slips and electrical hazards. Moreover, working under direct, intense sunlight can lead to heat-related illnesses.

- **Ladder Safety**: Always ensure that the ladder is on stable ground. If on a roof, secure it properly to prevent it from shifting. Also, maintain a 3-point contact (two hands and a foot or two feet and a hand) when climbing.

- **Know Your Limits**: DIY doesn't mean doing everything on your own. If there's a part of the installation you're unsure about, especially when it comes to electrical connections, consider consulting or hiring a professional for that segment.

The thrill of setting up a solar system yourself is undeniably fulfilling. However, it's essential to remember that no phase of the installation is worth compromising your safety. With the right gear and adherence to safety protocols, you can ensure that your journey into sustainable energy is not just successful but also safe.

CHAPTER 2: Step-by-Step Panel Installation

Mounting the Panels

The transformational moment in any solar project comes when the panels – the primary energy collectors – find their place, either upon a roof or a piece of land. It's the point where the theoretical merges with the practical. While the panels' technical nuances were discussed in earlier books, here, we venture into the tangible act of mounting them. The process, while not overwhelmingly complex, requires precision, understanding, and adherence to best practices.

Preliminary Steps: Getting Ready

Before any screws are turned or brackets are placed, there's groundwork to be done.

- **Choosing the Right Mounting System**: Depending on where you're planning to install the panels – on a roof, on the ground, or on a tracking system – you'd need different mounting systems. For roofs, flush mounts are common, while pole mounts are suitable for ground installations and tracking systems.

- **Determining the Solar Array's Layout**: This involves finalizing how many rows and columns of solar panels will be installed, keeping in mind the available space and the panels' size.

- **Site Cleaning**: Whether it's a rooftop or a piece of land, ensure the site is clean, devoid of debris, and, in the case of the ground, leveled.

Commencing the Installation

- **Marking & Measuring**: Begin by marking out the exact places where the mounting brackets will go. This is a critical step, as it ensures uniform spacing and alignment for the panels. When dealing with rooftops, you'd also want to ensure these brackets align with the rafters for maximum support.

- **Installing the Mounting System**: Once the markings are in place, start with the installation of the brackets or mounts. If it's a rooftop installation, drilling might be required to fix the brackets securely. Always ensure that any drilled holes are properly sealed to prevent potential leakages.

- **Setting up the Rails**: On these brackets, the next step is to fix the rails. These are the structures on which the solar panels will eventually sit. The rails need to be parallel, straight, and firmly attached to the mounting brackets.

- **Securing the Panels**: With the foundation ready, it's time for the panels. Carefully place each panel on the rails and secure them using the provided clips or bolts. While doing so, ensure that the panels are not overly tightened, as they need a little space for expansion and contraction due to temperature variations.

- **Grounding**: One crucial aspect of mounting solar panels is ensuring they are properly grounded. Grounding protects the system and the structure it's attached to from possible electrical faults. Attach the grounding wire to each panel and ensure it's connected to the grounding rod in the ground.

Considerations & Best Practices

- **Orientation and Angle**: For optimal energy capture, panels in the Northern Hemisphere should face true south, and those in the Southern Hemisphere should face true north. The tilt angle should be adjusted based on your latitude and the time of year to maximize sun exposure.

- **Avoiding Shadows**: Even a small shadow on one part of a solar panel can significantly reduce its efficiency. Ensure that no nearby structures, trees, or other obstacles cast shadows on the panels, especially during peak sunlight hours.

- **Ventilation**: Solar panels can get hot. Ensure there's enough gap between the roof and the panel (if roof-mounted) for air to flow and cool the panels. This not only boosts efficiency but also prolongs the panel's lifespan.

- **Future Maintenance**: While mounting, always keep in mind that you'd need to access these panels in the future for maintenance. Ensure there's enough space for a person to move and work around them.

Electrical Integration and Connection

Solar panel installation is more than just mounting panels in sunlight; it's about effectively integrating them into a building's electrical system. This phase of the process determines how efficiently the harvested sunlight gets converted into usable power. A properly integrated system ensures that the power generated is not just ample but also safe for all connected appliances and devices.

The Core Components of Solar Electrical Integration

There are several integral parts to this process:

1. **Solar Inverters**: These devices transform the direct current (DC) produced by the solar panels into alternating current (AC) used by most household devices and national grids. The choice of inverter, be it string inverters, microinverters, or power optimizers, can influence the efficiency of energy conversion.

2. **Wiring and Conduit**: These are the channels through which the electricity flows. Using the right type of wire, correctly gauged for the expected energy output, is critical to prevent energy losses and potential fire hazards.

3. **Safety Disconnects**: These switches allow the solar system to be manually disconnected, either for maintenance or in case of emergencies. They ensure that energy isn't flowing when it shouldn't be, providing an essential safety layer for workers and residents.

Detailed Integration Process

- **Inverter Installation**: Depending on the type, inverters can be mounted either near the solar panels or close to the main power system of the building. When selecting the installation spot, consider:

- **Accessibility**: It should be easily accessible for regular checks and maintenance.

- **Ventilation**: Inverters can produce heat, so it's crucial to ensure they have proper ventilation to avoid overheating.

- **Protection from Weather**: While many inverters are weatherproof, placing them under some form of shelter can prolong their life.

- **Wiring the Panels**: With the inverter in place, it's time to connect the panels. This process involves:

  - **Arrangement**: Solar panels produce DC, and often, they are connected in series to add up their voltages, forming a 'string'. Depending on the system's size, you may have multiple strings. Always follow the manufacturer's guidelines regarding the maximum number of panels per string.

  - **Routing the Cables**: The cables from these strings are then routed to the inverter. Use appropriate conduits to protect the wires from environmental factors. The length of the cable run should be minimized to reduce energy losses.

- **Integrating with the Building's Electrical System**: Once the panels are connected to the inverters, the next step involves integrating the system into the building's main power supply:

  - **Connection to the Grid**: For systems integrated with the national grid, a two-way meter is installed. This meter not only records the electricity consumed from the grid but also the excess power fed back into it.

- **Isolated Systems**: In off-grid setups, the electricity generated is often stored in batteries for later use. This requires the installation of charge controllers to ensure the batteries are charged optimally.

- **Safety Protocols**: Electricity, by its nature, is hazardous, and solar installations are no exception. Safety components include:

  - **Disconnect Switches**: Positioned between the solar setup and the building's main supply, these switches allow for a complete shutdown of the solar system.

  - **Surge Protectors**: Solar setups, especially those on open grounds, can be vulnerable to lightning strikes. Surge protectors help in shielding the system from such high-voltage spikes.

  - **Grounding**: Grounding all metallic parts of the solar system is crucial to prevent potential shocks or electrocution. It ensures that in the case of a fault, the stray electricity is directed safely into the earth.

Maintenance and Monitoring

Post-installation, regular checks and maintenance ensure optimal functioning:

- **Monitoring Systems**: Many modern solar setups come with monitoring systems, allowing homeowners to keep track of energy generation and consumption patterns. These systems can often detect if individual panels underperform, prompting timely maintenance.

- **Regular Inspections**: Schedule periodic inspections to check for any physical damages to the panels, wear and tear in wiring, or other issues. Components like inverters may also require firmware updates for optimal performance.

- **Cleaning**: Over time, panels can accumulate dust, pollen, or bird droppings. This can reduce their efficiency. Hence, periodic cleaning, using water and a soft brush, can help maintain their energy output.

The act of integrating solar panels electrically is where the rubber meets the road. It's where the clean energy harvested gets prepped to power our lives. A meticulous, safety-first approach ensures that this power doesn't just light up homes, but does so safely, efficiently, and sustainably. As we journey further, our next topic will shine light on the essential practices to protect and prolong the life of these solar installations.

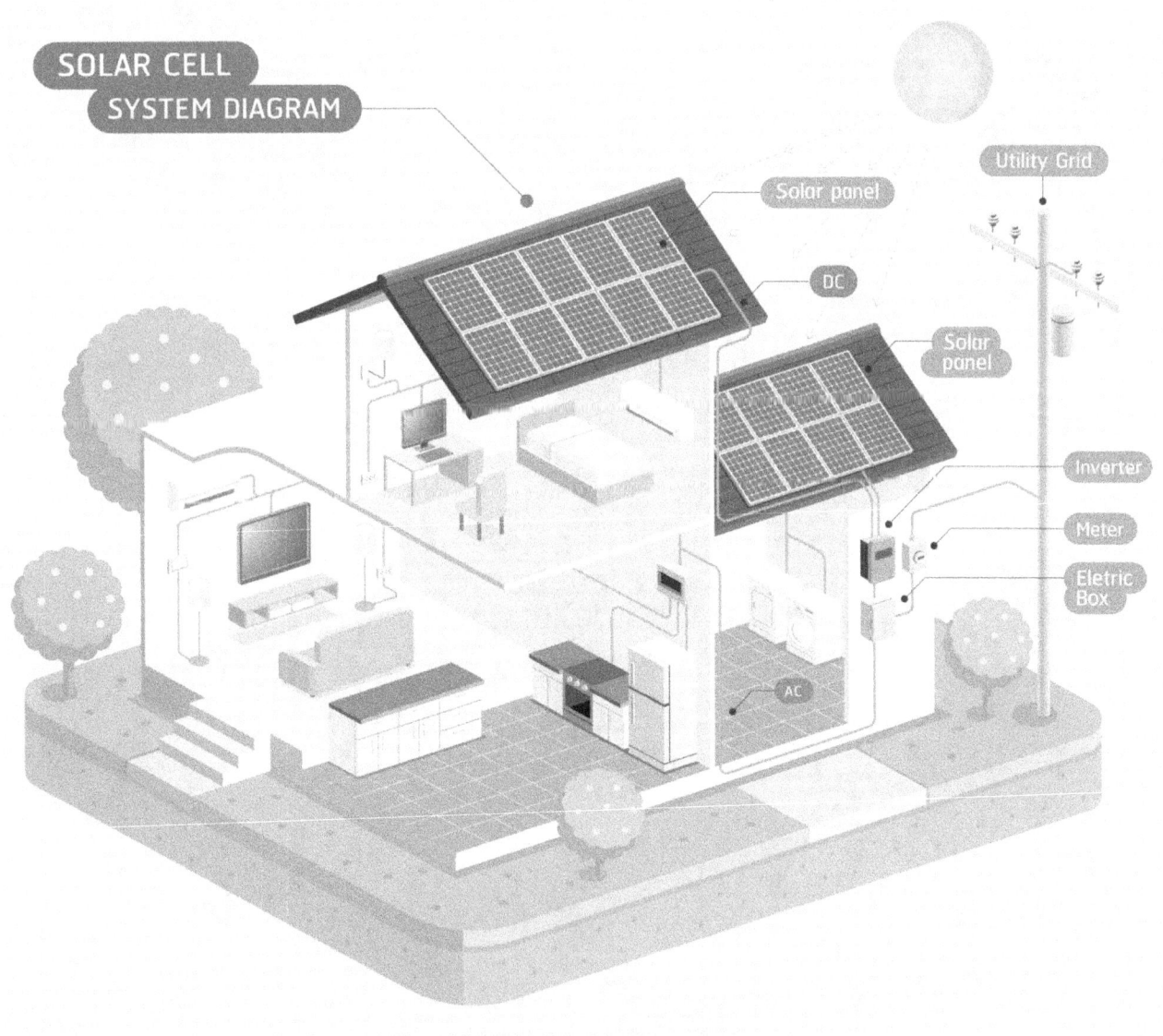

CHAPTER 3: Setting Up the Inverter and Batteries

Inverter Installation

At its core, every photovoltaic system thrives on the power of transformation. It's not just about harnessing the raw energy of the sun but converting it into a usable form for our homes. This is where the inverter shines as the silent hero. Acting as the bridge between raw energy and functional power, an inverter is indispensable.

Understanding the Role of the Inverter

The solar panels on your roof or property generate Direct Current (DC) power. However, our homes and most electrical appliances run on Alternating Current (AC) power. The inverter's primary role is to convert this DC power into AC power, making it usable for our daily needs.

Choosing the Right Inverter for Your Needs

The inverter is not a one-size-fits-all component. Its capacity and type should align with the energy requirements and specifications of your solar installation.

- **Size Matters**: An inverter's capacity should match the total wattage your solar panels produce. It's crucial to ensure that the inverter can handle the maximum power your solar system can generate.

- **Types of Inverters**: There are primarily three types of inverters used in residential solar installations:

  - **String Inverters**: Ideal for larger installations with panels facing the same direction, and with minimal shading issues.

- **Microinverters**: Perfect for homes with shading concerns or where panels face multiple directions. Here, each panel gets its own inverter, ensuring optimal performance.

- **Power Optimizers**: A middle ground between string inverters and microinverters, power optimizers are positioned at each panel but work with a centralized inverter. They ensure each panel performs at its peak while also benefiting from a consolidated inverter setup.

The Installation Process

Properly setting up your inverter ensures the seamless operation of your solar system. Here's a step-by-step guide to the installation process:

- **Selecting the Ideal Location**: The location of the inverter can influence its efficiency. While some inverters are designed for outdoor installations, others are better suited for indoor environments. Ensure it's placed in a cool, shaded spot, preferably close to the main power system to minimize energy loss.

- **Wall Mounting**: Ensure the wall chosen for mounting can handle the inverter's weight. Most inverters come with mounting brackets or a mounting system. After marking the appropriate points, use heavy-duty screws to affix the inverter, ensuring it's level and secure.

- **Making Electrical Connections**: Connect the DC inputs from the solar panels to the inverter. Ensure these connections are tight and secure. Also, connect the AC output from the inverter to your home's main electrical panel. Ensure you're using the right gauge of wires, and the connections are snug.

- **Grounding the Inverter**: Like all electrical equipment, inverters need grounding to prevent electrical shocks. Connect a grounding wire from the inverter to a grounding rod embedded in the earth.

- **Configuring the System**: Modern inverters come with digital displays and interfaces. Once powered on, set up the inverter by choosing the right configurations, usually specified in the user manual.

Safety First: Tips for a Safe Installation

- **Switch Off the Main Power**: Before starting any electrical work, always turn off the main power to prevent any accidental shocks.

- **Use Insulated Tools**: Using insulated tools ensures that you're not at risk of any inadvertent electrical contact.

- **Wear Safety Gear**: Even if you're working indoors, wear safety gloves and safety glasses. This protects you from potential sparks or wire fragments.

- **Keep the Area Dry**: Water and electricity are a dangerous combination. Ensure the installation area is dry, and avoid working during wet conditions.

- **Follow the User Manual**: Every inverter model is unique. Always refer to and follow the manufacturer's instructions for installation.

Post-installation Best Practices

After the installation, it's essential to routinely inspect your inverter for any signs of wear, damage, or malfunction. Listen for any unusual sounds, and periodically check for any error messages on the display.

Inverter technology has come a long way. Modern inverters often provide smart functionalities, allowing homeowners to monitor their solar system's performance in real-time. If your inverter has this feature, make the best use of it. By keeping an eye on your system's performance metrics, you can ensure everything operates smoothly, making necessary adjustments when required.

Battery Connections and Setup

If the inverter is the heart of a solar system, batteries are the lifeblood that ensures the system thrives even when the sun isn't shining. Batteries store the excess energy produced during sunny hours, making it available for use during nighttime, cloudy days, or periods of high demand. This ability to store and release energy as required makes solar batteries a cornerstone of any efficient photovoltaic system.

Decoding the Science Behind Solar Batteries

Solar batteries function on simple science: they store energy in a chemical form. When the solar panels produce more energy than the home can use, instead of sending it back to the grid, this energy is directed to the batteries. When the panels aren't producing enough energy, the batteries discharge, providing the necessary power to the home.

Choosing Your Solar Battery

There's a range of solar batteries in the market, each with its specifications and suited for different types of installations.

- **Lithium-ion Batteries**: These are the most common type of solar batteries. They're known for their longevity and high energy density, making them perfect for residential solar systems.

- **Lead-acid Batteries**: An older technology, they are often cheaper but have a shorter lifespan than lithium-ion batteries.

- **Flow Batteries**: These are a newer type of solar battery. They have a longer lifespan than both lithium-ion and lead-acid batteries but are typically larger and more expensive.

- **Saltwater Batteries**: Using saltwater as their electrolyte, these batteries are relatively new to the market. They're known for their safety and sustainability, as they don't contain heavy metals.

Setting Up Your Solar Battery: A Step-By-Step Guide

1. **Select the Ideal Spot**: Batteries function best in stable temperatures. Place them in a location that's neither too hot nor too cold. Some batteries are designed for outdoor installations, while others are for indoors. Choose a spot that's easily accessible for maintenance.

2. **Installation of Battery Enclosures**: Many batteries need protective enclosures. This ensures they're shielded from the elements and adds an extra layer of safety. Securely mount these enclosures following the manufacturer's guidelines.

3. **Connect to the Inverter**: Your battery will connect to the solar system through an inverter. Some systems have battery-specific inverters, while others use hybrid inverters for both panels and batteries. Ensure you've got the right setup for your system.

4. **Wire the Battery**: Using the appropriate gauge of wire, connect the battery to the inverter. Ensure all connections are secure. Loose connections can reduce efficiency and be potential fire hazards.

5. **Install a Battery Management System (BMS)**: This system monitors the cells within the battery, ensuring they charge and discharge uniformly. A BMS can greatly enhance the longevity of your battery.

6. **Configure the System**: Modern batteries come with digital controls allowing homeowners to specify when they want to draw power from the batteries, and when they want to charge them. Set these according to your energy needs.

Safety Protocols to Keep in Mind

- **Regular Inspection**: Periodically inspect your batteries for signs of wear, corrosion, or leakage. Any of these can impair battery performance and may be potential safety hazards.

- **Avoid Overcharging**: Overcharging can damage batteries and reduce their lifespan. Ensure your system prevents this from happening.

- **Keep Batteries Clean**: Dust and debris can impact battery performance. A simple wipe-down can often suffice to keep them in top shape.

- **Follow Manufacturer's Guidelines**: Each battery type and brand can have specific installation and maintenance instructions. Always adhere to these for optimal performance and safety.

The Role of Batteries in a Solar Future

As the world moves toward a more sustainable energy paradigm, the role of storage becomes paramount. And in this landscape, batteries stand as silent sentinels, ensuring homes remain powered even when the sun sets. With technological advancements, we're witnessing batteries that are more efficient, longer-lasting, and more affordable. Investing in a good solar battery ensures that your solar journey is smooth, efficient, and truly self-reliant.

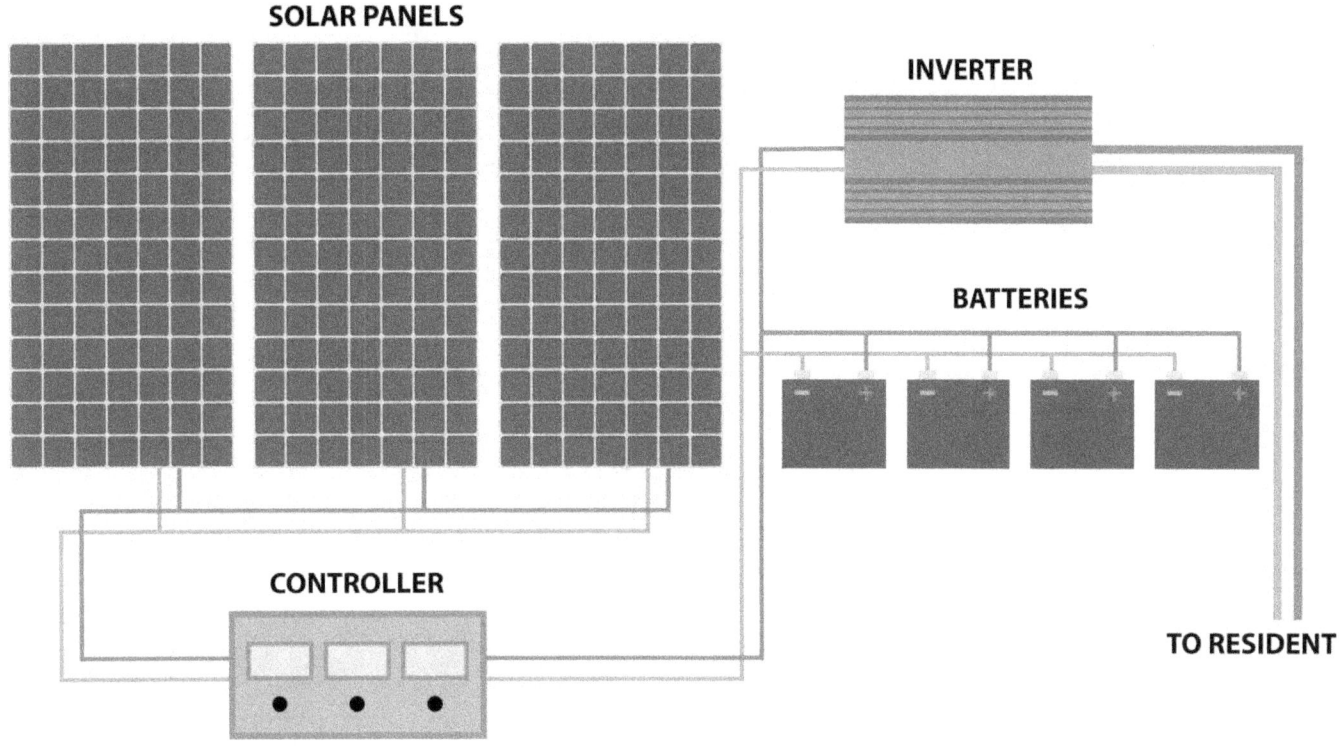

SOLAR PANELS

INVERTER

BATTERIES

CONTROLLER

TO RESIDENT

CHAPTER 4: System Testing and Commissioning

Initial System Checks

After installing a solar system, ensuring its optimal performance becomes the next imperative step. Before a system is officially commissioned and begins its long-term service of converting sunlight into electricity, it undergoes a series of initial system checks. These checks not only guarantee the safety of the system but also validate its efficiency and effectiveness.

Understanding the Importance of Initial Checks

Diving right into the operation of a newly installed solar system without proper checks can lead to underperformance, potential safety hazards, and long-term damage. The initial system checks act as a "health-checkup" for the solar installation, ensuring all parts are functioning as they should and the system is primed to deliver its best performance.

Physical Inspection: The First Step

Before any technical assessments, a thorough physical inspection is essential.

- **Panels**: Inspect for any visible damage or dirt on the panels. Even a slight shadow or dirt can reduce the efficiency of the panel.

- **Wiring**: Check the physical state of all wires, ensuring they are well-protected, insulated, and show no signs of wear or damage.

- **Mounts and Racks**: Confirm that all mounts and racks are securely fastened and provide stable support to the panels.

Voltage and Current Testing

Using a multimeter, you'll want to measure the open-circuit voltage and short-circuit current. These measurements should closely align with the manufacturer's specifications. Significant deviations could indicate issues like damaged panels or wiring faults.

Inspecting the Inverter Display

Modern inverters come with digital displays that provide real-time data on the system's performance.

- **Operational Status**: Ensure the inverter indicates it's operational and isn't displaying any fault messages.

- **Power Output**: Cross-reference the current power output with the expected output for the given sunlight conditions. It should be within a reasonable range.

- **Energy Produced**: The inverter will show the total energy produced since its installation. Monitor this over a few sunny days to ensure it aligns with expected projections.

Functional Tests of Protection Devices

Every solar system comes equipped with various protection devices, such as fuses, circuit breakers, and disconnect switches. It's vital to ensure each of these is functioning correctly.

- **Isolation Test**: This test ensures that there's no unwanted electrical connection between the solar system's conductive parts and the earth, which could lead to unwanted power flows or even electrical shocks.

- **Continuity Test**: Here, we check the effectiveness of the connections. It ensures that the current flows smoothly and uniformly throughout the system.

Commissioning the System

Once all tests are passed, and any identified issues are rectified, the system is ready to be officially commissioned. This involves:

- **Documentation**: Maintain a detailed record of all tests performed, their results, and any corrective actions taken. This document becomes a vital reference for future maintenance or troubleshooting.

- **System Activation**: With everything in place, switch on the system and allow it to start feeding power either to your home, the batteries, or back to the grid.

Monitoring and Maintenance

Every sunray captured and converted adds up in our global shift towards renewable energy. Monitoring helps track the performance of the solar system, ensuring each ray is utilized to its fullest potential.

- **Performance Metrics**: With monitoring systems in place, homeowners can gauge the real-time performance of their solar installation. These metrics offer insights into the energy produced, peak performance times, and any discrepancies in expected outputs.

- **Predictive Analysis**: Modern solar monitoring systems can predict potential issues before they escalate. For instance, if a particular section of the panel consistently underperforms, it could be due to shading or an impending malfunction.

- **Energy Consumption Tracking**: Beyond production, monitoring tools can also provide a detailed analysis of energy consumption patterns. This can guide homeowners to optimize their energy usage, ensuring they make the most of the solar energy they produce.

Maintenance: The Guardian of Solar Longevity

While solar systems are designed for durability, periodic maintenance ensures they continue to operate at their prime.

- **Routine Cleaning**: Dust, pollen, bird droppings, and other debris can accumulate on the panels over time. Periodic cleaning ensures that these obstructions don't hamper the solar absorption efficiency. However, it's essential to use the right cleaning methods to avoid scratching or damaging the panels.

- **Visual Inspections**: Over time, weather conditions can strain the physical components of the system. Regular visual checks can help spot issues like loosening mounts, frayed wires, or even critter nests that might have formed underneath panels.

- **Professional Check-ups**: While many aspects of solar maintenance can be DIY, it's beneficial to have professionals conduct a comprehensive inspection annually. They can identify and rectify nuanced issues that might go unnoticed otherwise.

Leveraging Technology for Proactive Maintenance

In an age of smart homes and IoT, solar systems aren't left behind. Many solar installations now come equipped with advanced sensors and software solutions.

- **Automated Alerts**: These systems can send real-time notifications if they detect inefficiencies or malfunctions. It allows homeowners to address issues promptly.

- **Remote Troubleshooting**: Some advanced monitoring systems offer remote troubleshooting. Solar experts can often rectify software-related issues or guide homeowners through specific maintenance tasks without the need for a physical visit.

- **Integrating with Smart Homes**: Many monitoring systems can integrate seamlessly with other smart home systems. This can enable homeowners to synchronize their energy consumption patterns with solar energy production, optimizing usage.

The Resonance of Consistency

As with any sophisticated system, consistency is the heartbeat of efficiency. Regular monitoring and periodic maintenance form a rhythm that keeps the solar system performing at its peak. It's this unwavering attention to detail that ensures the longevity of the system, maximizes ROI, and fortifies the commitment to a sustainable future.

BOOK 5

Advanced Solar Solutions And Innovations

CHAPTER 1: Bifacial Solar Panels

How Bifacial Panels Work

When we think of solar panels, often, a monochrome palette of blue or black panels against rooftops or vast fields comes to mind. But as technology advances, the world of solar energy is witnessing innovative solutions that challenge the conventional. Enter bifacial solar panels, a game-changer that harnesses the sun's energy not from one, but two sides. Let's dive deep into understanding how these panels work and why they might just represent the future of solar energy.

The Anatomy of Bifacial Panels

At their core, bifacial solar panels aren't dramatically different from their monofacial counterparts. Like traditional panels, they consist of silicon cells responsible for converting sunlight into electricity. However, the difference lies in the fact that these cells are encapsulated in a transparent back sheet or glass, allowing them to capture sunlight from both the front and the back.

The Magic of Dual Capture

Sunlight, as we know, doesn't just fall directly onto solar panels. It reflects, refracts, and dances around, especially on surfaces beneath the panels like rooftops, pavements, or the ground. These surfaces often reflect what's known as 'albedo' light. Bifacial panels, with their transparent backing, are uniquely positioned to capture this albedo light, turning reflections into productive energy.

Boosted Efficiency: A Play of Light

One might wonder how capturing light from the rear makes a significant difference. After all, isn't the most substantial sunlight coming from above? While it's true that direct sunlight is potent, the additional albedo light can increase energy yield by a notable margin. Depending on the reflecting surface and installation conditions, bifacial panels can boost energy output by 5% to 30%.

Bifacial Technology: Silicon and Beyond

While bifacial technology can be incorporated into various types of solar cells, it's most commonly associated with monocrystalline and polycrystalline silicon cells. However, recent innovations are also integrating bifacial designs into thin-film solar technologies, broadening the spectrum of how and where bifacial panels can be used.

The Role of Installation

The efficacy of bifacial panels is intrinsically linked to their installation. Elevated installations, where panels are positioned higher above the ground or other reflecting surfaces, often yield better results. This is because a greater distance allows more reflected light to reach the rear side of the panels. Additionally, the type of surface beneath matters. Light-colored or reflective surfaces, such as white roofs or concrete, can significantly enhance the amount of albedo light captured.

Symbiosis with Tracking Systems

Bifacial panels often find their perfect partner in solar tracking systems. Trackers adjust the angle of solar panels throughout the day to follow the sun's path. When bifacial panels are mounted on trackers, they can capture direct sunlight more efficiently from the front while still harnessing the albedo light from the rear, supercharging their energy yield.

The Path Ahead: Constant Evolution

Like all technology, bifacial panels are continually evolving. As research advances, we're seeing enhancements in cell designs, transparency materials, and installation techniques, each pushing the boundaries of what's possible with bifacial technology.

Benefits and Implementation of Bifacial Panels

Bifacial solar panels, by virtue of their dual-sided energy capture, offer a dynamic shift in the renewable energy landscape. But what are the tangible benefits they bring, and how can one implement them optimally? This section delves into the advantages and practical considerations surrounding bifacial panels.

The Dual Advantage: Enhanced Energy Yields

One of the most apparent benefits of bifacial panels is the increase in energy production. By capturing albedo light — the sunlight reflected off surfaces below the panels — they can enhance energy yield significantly. Depending on the installation, surroundings, and albedo of the ground, energy output can surge anywhere from 5% to 30% compared to monofacial panels.

Durability and Longevity: Built to Last

Bifacial panels, often constructed with glass-on-glass designs, are inherently more durable than traditional panels. This double glass construction not only ensures better protection against environmental factors like humidity and temperature fluctuations but also offers resistance against mechanical stresses, ensuring a longer lifespan.

Reduced Degradation: Staying Efficient Longer

Solar panels, over time, undergo degradation, which means their efficiency in converting sunlight into electricity slightly diminishes every year. Bifacial panels have shown lower degradation rates compared to their monofacial counterparts, ensuring that they retain a higher percentage of their original efficiency over the years.

Versatile Installation: Adapting to the Environment

Bifacial panels provide greater flexibility when it comes to installation. Whether they're installed over light-colored rooftops in urban settings or amidst green fields in rural areas, they adapt by harnessing the reflected sunlight.

This versatility means they can be implemented in varied environments, maximizing their energy capture potential.

Cost-Effectiveness Over Time

While the initial investment for bifacial panels might be slightly higher than traditional panels, the enhanced energy yields, combined with their durability and reduced degradation, often result in a better return on investment in the long run.

Practical Steps for Bifacial Panel Implementation

1. **Assessment of Location**: Before installing bifacial panels, assess the location's albedo. Light-colored or reflective surfaces enhance the performance of bifacial panels. Consider whether the location is suitable or if changes to the ground or surrounding surfaces can be made to boost albedo.

2. **Elevated Installation**: To maximize the capture of reflected light, bifacial panels should be installed at a height. This allows more reflected sunlight to reach the rear side.

3. **Integrate with Tracking Systems**: To further enhance energy yields, consider integrating bifacial panels with solar tracking systems. These systems will adjust the panel angle throughout the day, ensuring optimal sunlight capture.

4. **Monitor and Maintain**: After installation, continuously monitor the performance of bifacial panels. Regular maintenance, like cleaning and checking for any obstructions, ensures they operate at peak efficiency.

5. **Engage with Professionals**: Given the unique nature of bifacial panels, it's crucial to engage with professionals who have experience in their installation and maintenance. Their expertise can guide optimal placement and ensure the panels are working at their best.

BIFACIAL
SOLAR PANELS

THE PANEL GETS HIT BY SUNLIGHT DIRECTLY AND ALSO BY LIGHT THAT BOUNCES OFF THE GROUND

DIFFUSED SUNLIGHT, ALONG WITH LIGHT THAT'S REFLECTED FROM THE GROUND, TOUCHES BOTH THE FRONT AND BACK OF THE PANEL

CHAPTER 2: Smart Solar Systems

Integrating IoT with Solar

The Internet of Things (IoT) has dramatically transformed industries and daily life. By connecting everyday devices and systems to the internet, IoT promises improved efficiency, data-driven insights, and remote management. So, when this cutting-edge technology is married to solar energy, what wonders can we expect?

From Reactive to Proactive: The IoT-enhanced Solar System

IoT integration allows solar systems to transition from merely reactive setups — that generate power when the sun shines — to proactive, intelligent systems. These are systems that anticipate, adjust, and even communicate to ensure optimal performance.

Real-time Monitoring and Analytics

Solar systems have always generated data. But with IoT, the real magic lies in how this data is collected, analyzed, and utilized. IoT devices can continually monitor a solar panel's performance, analyzing data in real-time, and detecting inefficiencies or malfunctions instantaneously. By identifying any issues immediately, corrective action can be taken swiftly, ensuring minimal disruption.

Predictive Maintenance Through Data Insights

Traditional solar maintenance is often scheduled or arises when an obvious problem occurs. IoT transforms this by facilitating predictive maintenance. By analyzing data trends and using machine learning algorithms, IoT can predict when a component is likely to fail or when efficiency might drop, allowing for preemptive maintenance. This not only prolongs the system's life but also ensures consistent power generation.

Remote Configuration and Control

With solar systems connected to the internet, remote management becomes a reality. Whether it's adjusting the angle of solar trackers or modifying inverter settings, operators can make changes from anywhere in the world using a simple internet-connected device. This flexibility is invaluable, especially for large-scale installations or installations in remote areas.

Optimizing Energy Consumption

Beyond the generation side, IoT can also impact how solar energy is consumed. By integrating IoT with home or industrial automation systems, devices can use energy more intelligently. For instance, during peak solar generation times, IoT systems could prioritize high-energy tasks, like charging electric vehicles or running industrial machinery.

Grid Integration and Energy Trading

As decentralized solar systems become more common, integrating these with the main grid becomes complex. IoT can help manage this by analyzing when to feed energy into the grid or when to draw from it, based on real-time energy prices, demand, and supply factors. Additionally, in future energy markets, homeowners could even sell excess energy, with IoT devices facilitating these transactions automatically.

Implementation Steps: Making Solar Smart with IoT

1. **Evaluation**: Before embarking on IoT integration, evaluate the current solar setup. Identify what data points would be valuable, and determine the areas where IoT could bring about significant improvements.

2. **Choose Reliable IoT Devices**: The market is saturated with IoT devices, so select those known for durability, security, and compatibility with existing solar components.

3. **Integrate with Data Analysis Platforms**: Gathering data is one thing, but analyzing it effectively is another. Utilize platforms designed to provide actionable insights from the data harvested by IoT devices.

4. **Prioritize Security**: As with all internet-connected systems, security is paramount. Ensure all devices are secure, regularly updated, and protected against potential cyber threats.

5. **Engage with IoT and Solar Experts**: This integration is a marriage of two technical fields. Engage with experts who understand both solar systems and IoT to ensure seamless integration and optimal performance.

With the convergence of solar energy and IoT, we stand on the brink of a revolution in renewable energy management. These interconnected systems hold the promise of making solar energy more efficient, more responsive, and even more integrated into our daily lives.

Advanced Monitoring and Analytics

In an era where every byte of data matters, understanding and interpreting the vast amount of information generated by solar systems is pivotal. Solar analytics, especially in recent times, has matured beyond mere performance monitoring. The contemporary solar landscape thrives on advanced analytics, which delve deeper, offering insights that can transform the efficiency, lifespan, and overall profitability of solar installations.

Unraveling the Data Matrix: What Advanced Analytics Offers

The power of advanced analytics in the solar sector can be likened to peering through a magnifying glass, revealing patterns, anomalies, and opportunities that were previously indiscernible.

Granular Data Analysis

Traditional monitoring might tell you that a solar panel's efficiency has dropped. In contrast, advanced analytics provides a comprehensive breakdown — from the performance of individual solar cells within a panel to the impact of accumulated dust or shade. Such granularity allows for targeted interventions, which can significantly enhance the system's output and lifespan.

Predictive Performance Analysis

Harnessing machine learning and sophisticated algorithms, these analytic tools can predict future system performance based on historical data, environmental factors, and emerging trends. This foresight can be invaluable in planning maintenance, managing energy consumption, and ensuring that the system continually operates near its peak efficiency.

Visual Representations

Advanced analytics platforms often employ detailed visualizations, from heat maps highlighting underperforming panels to 3D simulations showing the impact of potential obstructions like future building projects or tree growth. Such visual tools can assist stakeholders in understanding complex data sets at a glance and making informed decisions.

Environmental and Geographic Analytics

By integrating local environmental and geographic data, these platforms can provide insights on how local weather patterns, seasonal changes, or even microclimates can influence solar performance. For example, understanding the frequency of cloud cover, morning fog, or seasonal dust can aid in predictive analytics and system design optimization.

Financial and ROI Analytics

Beyond technical performance, advanced monitoring platforms also provide detailed financial analytics. These can predict future energy savings, calculate return on investment over varying time frames, and even forecast revenue for systems feeding energy back into the grid. Such financial insights are invaluable for both individual homeowners and large-scale commercial installations, particularly when considering expansion or new investments.

Integration with Broader Smart Systems

Many advanced analytics tools are designed to integrate seamlessly with broader building management systems, smart home setups, or Industrial automation systems. This holistic approach means that solar data can influence and be influenced by other systems, leading to a more integrated, efficient energy ecosystem.

Adopting Advanced Monitoring: Steps to Leverage Deep Insights

1. **Research and Selection**: Not all advanced analytics platforms are made equal. It's essential to research and select a platform that aligns with the specific needs of the solar installation, whether it's a small residential setup or a sprawling commercial solar farm.

2. **Training and Adaptation**: Advanced tools often come with a steeper learning curve. Investing time in training — for homeowners, technicians, or system managers — ensures that the platform's full capabilities are leveraged.

3. **Regular Data Reviews**: The value of analytics is fully realized when data is regularly reviewed and acted upon. Whether it's adjusting system configurations, planning maintenance, or redesigning a part of the solar array, data-driven decisions can markedly improve performance.

4. **Staying Updated**: The field of solar analytics is continually evolving. Regular software updates, staying informed about new analytic methodologies, and even attending industry seminars can ensure that the system remains at the forefront of analytic capabilities.

Advanced monitoring and analytics act as the guiding light, ensuring that every solar installation — regardless of its scale — operates efficiently, profitably, and sustainably.

CHAPTER 3: Solar Plus Storage Solutions

The Rise of Home Energy Storage

The contemporary energy dialogue is increasingly focusing on decentralized energy storage – particularly in homes. The transition from large-scale grid dependency to self-sustained households is not merely a trend; it's a revolution.

Home Energy Storage: An Overview

At its core, home energy storage allows homeowners to store energy, usually generated from renewable sources like solar panels, for later use. This offers numerous benefits, from increased energy independence and resilience against power outages to the potential of monetizing excess energy.

Why Home Energy Storage is Gaining Momentum

1. Solar Surge and the Need for Storage: With the increase in residential solar installations, there's been a simultaneous rise in the demand to harness excess solar energy. On days when the sun is shining brightly, and energy production exceeds consumption, having a storage solution becomes not just useful, but necessary.

2. Enhanced Energy Independence: Relying solely on the grid can be a gamble, especially in areas prone to power outages or with unreliable energy infrastructures. Home energy storage acts as a buffer, ensuring power availability even during grid failures.

3. Economic Incentives: As renewable energy becomes cheaper, storing it also becomes economically viable. In some regions, energy prices fluctuate throughout the day, allowing homeowners to consume their stored energy during peak rates and save money.

4. Environmental Considerations: Storing and using renewable energy at home reduces the dependence on fossil fuel-based grid power, aligning with global efforts to combat climate change.

5. Technological Advancements: Modern energy storage solutions, especially batteries, have seen leaps in efficiency, longevity, and affordability. These innovations make home energy storage an attainable goal for many.

Diverse Energy Storage Solutions

While batteries, particularly lithium-ion, dominate the conversation, several other technologies are emerging, offering unique advantages:

- **Thermal Storage:** This system stores excess energy as heat in insulated repositories, like water tanks or specialized materials. The stored heat can be converted back into electricity or used for home heating.

- **Flywheel Storage:** Flywheels store energy kinetically. When there's an excess of energy, it's used to accelerate a rotor in a low-friction environment. When energy is required, the rotor's deceleration is converted back to electricity.

- **Compressed Air Energy Storage (CAES):** This involves storing energy by compressing air in underground reservoirs. When energy is needed, the compressed air is heated and expanded, driving a turbine to produce electricity.

The Path Forward: Integration and Optimization

Home energy storage isn't merely about installing a battery and forgetting it. It's about integrating storage into the broader home energy ecosystem. Smart technologies now allow homeowners to optimize when they draw from or feed into their storage, depending on various factors like weather predictions, planned energy consumption, and grid energy prices.

Moreover, community energy storage solutions, where a group of homes collectively invests in and benefits from a larger storage system, are emerging as feasible models, especially in urban settings.

Pairing Solar with Battery Systems

In the realm of renewable energy, solar panels and battery storage systems are often seen as two sides of the same coin. Their relationship is symbiotic: while solar panels harness the sun's energy, batteries ensure this energy is available whenever it's needed, creating a seamless, efficient energy ecosystem for homeowners.

<u>Why Pairing Makes Sense</u>

Harnessing the Inconsistency of the Sun: Solar panels, by nature, rely on sunlight, which is inherently inconsistent due to night-time, weather variations, and seasonal changes. Battery storage compensates for this inconsistency by storing excess energy produced during peak sunlight hours, making it available during periods of low or no sun.

Enhanced Reliability: Grid failures or blackouts can be detrimental, especially in areas prone to natural disasters or with aging infrastructure. When solar panels are paired with storage, homes can function off-grid, ensuring continuous power supply regardless of external disruptions.

Economic Efficiency: Time-of-use rate structures, where electricity prices vary depending on demand, can be navigated efficiently with battery storage. By utilizing stored energy during peak pricing times and drawing from the grid during off-peak hours, homeowners can optimize their energy costs.

Decentralizing Energy: Combining solar and storage not only empowers individual households but also decentralizes energy production and distribution. This reduces strain on central grids and paves the way for more resilient community-based energy networks.

Understanding the Mechanics

When solar panels produce more energy than the home is consuming, instead of feeding it back into the grid or letting it go to waste, this energy is directed towards charging the batteries. Advanced systems can even be programmed to prioritize battery charging, ensuring that storage is always maximized.

Once the batteries are fully charged, excess energy can then be fed back into the grid (if grid-tied) or used for other supplementary systems in the home. During non-productive solar hours or periods of high demand, the stored energy in the batteries is then utilized.

Choosing the Right Battery for Solar Systems

Not all batteries are created equal, and the choice largely depends on the homeowner's specific needs:

- **Capacity & Power:** Capacity refers to the total amount of electricity a battery can store, measured in kilowatt-hours (kWh). Power, on the other hand, determines how much energy a battery can provide at a given moment. A high-capacity battery could deliver a low power over a long duration or high power for a short time.

- **Depth of Discharge (DoD):** Most batteries need to retain some charge, and the DoD denotes the percentage of the battery's capacity that has been used. For instance, a 10 kWh battery with a DoD of 90% provides 9 kWh of usable capacity.

- **Battery Life & Warranty:** Given the investment, understanding the lifespan of a battery and the terms of its warranty is crucial. A common warranty might guarantee 5,000 cycles or ten years with at least 70% capacity retention.

Integrating smart systems with solar and battery setups offers adaptive solutions. These systems can predict weather patterns to optimize charging, adjust to real-time electricity pricing, or even integrate with home automation for demand-side energy management. Essentially, smart systems ensure the most efficient and cost-effective use of the energy harnessed and stored.

CHAPTER 4: Emerging Technologies

Perovskite Solar Cells

The continuous quest for more efficient, cost-effective, and flexible solar technology has led to the advent of perovskite solar cells, promising a revolutionary leap in photovoltaic technology. Named after the mineral perovskite, which they closely resemble in structure, these solar cells have been making headlines for their rapid efficiency improvements and unique properties.

Unraveling the Perovskite Magic

At the heart of this innovation is the perovskite material – a crystalline structure formed by combining organic molecules with metal halides. This concoction results in some remarkable features:

1. **High Light Absorption:** Perovskites have an innate ability to absorb visible light, which means they can generate a high amount of electricity even with a slim layer of material.

2. **Tunability:** One of the most promising attributes of perovskite solar cells is their tunability. By tweaking their composition, the absorption spectrum of these cells can be adjusted to capture more sunlight, allowing for multi-junction cells that can absorb a broader range of solar spectrum.

3. **Flexibility & Lightweight:** Unlike the rigid nature of traditional silicon solar cells, perovskite cells can be fabricated on flexible substrates, making them ideal for a wide array of applications, from wearable electronics to aerospace.

Efficiency Milestones

What's particularly striking about perovskite solar cells is their meteoric rise in efficiency. Within a decade, their lab-scale efficiencies have soared from a mere 3-

4% to over 25%, rivaling and even surpassing conventional silicon-based solar cells.

Manufacturing Advantages

The production of perovskite cells carries several advantages:

- **Low-Temperature Processing:** Unlike silicon cells, which require high-temperature processing, perovskite cells can be manufactured at much lower temperatures, making production less energy-intensive and potentially cheaper.

- **Solution-Based Fabrication:** Perovskites can be dissolved into a liquid and then printed onto a substrate, allowing for roll-to-roll manufacturing, akin to printing newspapers. This could significantly ramp up production speeds and reduce costs.

Challenges to Overcome

Despite the immense potential, perovskite solar cells are not without challenges:

- **Stability & Longevity:** Currently, one of the major roadblocks is the long-term stability of these cells. Exposure to moisture, heat, and UV light can degrade the perovskite material.

- **Toxicity Concerns:** Some perovskite cells use lead, raising environmental and health concerns. However, researchers are keenly exploring lead-free alternatives to mitigate this issue.

- **Scaling Up:** While lab-scale cells have shown impressive efficiency, scaling up to commercial-sized panels without efficiency drop-off remains a hurdle.

Promising Applications

The versatility of perovskite cells opens doors to applications previously unimagined:

- **Building Integrated Photovoltaics (BIPV):** Their flexibility and semi-transparency make them perfect for integrating into building facades, windows, or even roofs.

- **Portable Electronics:** Given their lightweight and flexibility, they could be integrated into backpacks, wearables, or even tents to provide on-the-go power.

- **Aerospace:** With weight being a premium in aerospace, the lightweight nature of perovskite cells could see them being used in satellites or high-altitude drones.

Perovskite solar cells, with their unique set of advantages, have the potential to reshape the solar industry. While challenges remain, the momentum in research and development is undeniable.

PEROVSKITE SOLAR CELLS

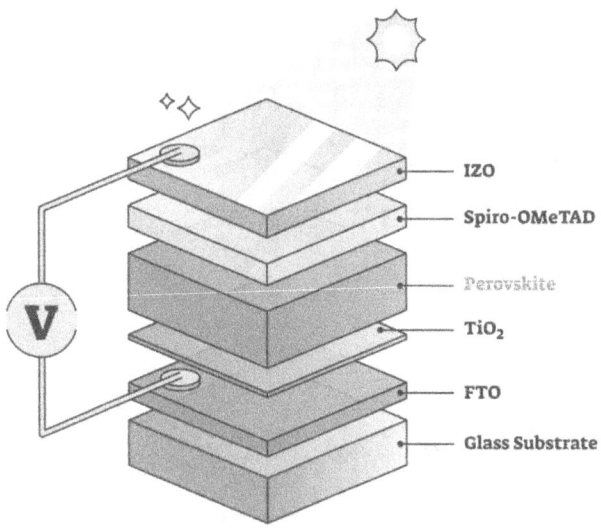

IZO

Spiro-OMeTAD

Perovskite

TiO$_2$

FTO

Glass Substrate

Transparent Solar Panels

Imagine a bustling city where every skyscraper's window doubles as a solar panel, where cars glide by with sun-harvesting sunroofs, and where everyday electronic gadgets are powered by light passing through their screens. This isn't a sci-fi vision of the distant future—it's the promise brought to us by transparent solar panels.

The Science of Transparency in Solar Panels

The traditional solar panel is opaque because it needs to absorb sunlight to convert it into electricity. Making a solar panel transparent seems counterintuitive at first—how can a panel generate power if it lets light pass through? The magic lies in the science of selectively harnessing parts of the solar spectrum.

1. **Harnessing the Invisible Spectrum:** Transparent solar panels primarily capture ultraviolet (UV) and infrared (IR) light—parts of the spectrum that are invisible to the human eye. While visible light passes through, these invisible wavelengths are absorbed and converted into electricity.

2. **Organic Photovoltaic (OPV) Cells:** A breakthrough in the realm of transparent panels is the development of organic photovoltaic cells. These cells use organic compounds that are fine-tuned to absorb only specific wavelengths of light, allowing the rest to pass through.

3. **Layered Structure:** To achieve transparency, these panels are typically designed with multiple ultra-thin layers. These layers not only facilitate the absorption of UV and IR light but also ensure structural integrity and durability.

Advantages Beyond Transparency

The allure of transparent solar panels extends beyond their see-through nature:

- **Aesthetic Integration:** Their transparent nature allows them to be seamlessly integrated into architectural designs without the aesthetic disruption typically associated with solar installations.

- **Urban Deployment:** Space is at a premium in urban settings. By turning windows and facades into energy generators, transparent solar panels can harness vast solar potential without consuming additional real estate.

- **Versatility:** From car sunroofs, smartphone screens, to bus stop shelters, the potential applications are boundless.

Challenges on the Horizon

Like all emerging technologies, transparent solar panels face hurdles:

- **Efficiency Trade-off:** Currently, transparent solar panels have a lower efficiency compared to their opaque counterparts. This is because they're not capturing the full spectrum of sunlight.

- **Cost Implications:** The advanced materials and manufacturing processes involved can make transparent panels more expensive than traditional panels, at least initlally.

- **Durability:** Given their proposed integration into daily use items like windows or screens, ensuring these panels remain effective and scratch-free over time is crucial.

Promising Implementations

The applications of transparent solar panels extend far beyond just skyscraper windows:

- **Consumer Electronics:** Devices like smartphones, smartwatches, and tablets could benefit from transparent solar panels integrated into their screens, providing auxiliary power and longer battery life.

- **Transportation:** Vehicles with sunroofs or large windshields can integrate these panels, offering an additional power source, especially beneficial for electric vehicles.

- **Public Infrastructure:** Bus shelters, pedestrian walkways, and public buildings can use transparent solar panels to power lights, displays, and other amenities.

Transparent solar panels represent the intersection of aesthetics and functionality. While there are hurdles to overcome, the potential implications for urban planning, architectural design, and daily life are profound.

BOOK 6

Off-Grid Living Embracing Independence

CHAPTER 1: The Allure of Off-grid Life

Breaking Free from Dependencies

In an increasingly interconnected world, where every aspect of life seems tethered to modern infrastructures, the idea of living off-grid has taken on a certain romantic allure. The notion of cutting ties, not just from the daily grind but from the very grid that powers it, speaks to a deep-seated desire for independence, autonomy, and self-sufficiency.

The Chains of Modern Dependency

We live in a society built upon dependencies. These aren't merely the benign interdependencies of community or family, but more complex ties that bind us to centralized services, corporations, and sometimes, even unseen global forces. Consider:

- **Energy:** Most households rely heavily on the centralized power grid, leaving them vulnerable to outages, price hikes, and environmental consequences.

- **Water:** Clean water, piped directly into homes, is a convenience many take for granted. Yet, this too is a form of dependency on public utilities and infrastructure.

- **Food:** The journey of the food on our plates often spans thousands of miles, involving numerous intermediaries. Our sustenance is bound to a global web of supply chains.

- **Communications:** From cellular networks to broadband internet, our means of communication are overwhelmingly centralized, often controlled by a handful of corporations.

The Off-Grid Philosophy: Reclaiming Autonomy

At its core, off-grid living is more than just a residential choice. It's a philosophy and a lifestyle. It's about:

- **Self-reliance:** Being capable of meeting one's basic needs, from energy to food, without relying on external systems.

- **Resilience:** Developing the ability to adapt and thrive, even when those external systems falter or fail.

- **Minimalism:** Recognizing the difference between wants and needs, and prioritizing the latter, often leading to a simpler, less cluttered life.

Steps Toward Independence

Breaking free from these dependencies doesn't necessarily mean isolating oneself from the modern world. It's about creating choices. Here's how many achieve it:

- **Energy:** Harnessing renewable sources like solar, wind, or hydroelectric power to generate electricity, thereby reducing reliance on the grid.

- **Water:** Implementing rainwater harvesting systems, digging wells, or utilizing natural springs to source water. Purifying and recycling water becomes a personal responsibility.

- **Food:** Engaging in agriculture, be it large-scale farming or small-scale gardening, raising livestock, or foraging, ensuring a fresher and more direct source of food.

- **Communication:** While some off-gridders choose to cut ties entirely, others opt for satellite communications or ham radio, ensuring connectivity on their own terms.

Beyond the tangible benefits of off-grid living, there are profound psychological rewards. The very act of taking control, of being the master of one's domain, instills a sense of confidence and peace. Every drop of water consumed, every morsel of food eaten, and every watt of electricity used becomes a testament to one's capabilities, resourcefulness, and resilience.

Environmental and Personal Benefits

Off-grid living is more than just a stylistic or economic choice. It is a testament to a holistic approach that intertwines personal well-being with environmental responsibility. Embracing this lifestyle brings forth numerous environmental and personal benefits that can transform our perception of modern living.

Harmony with Nature

Living off-grid often means being closer to nature. The daily routines of an off-grid life — from collecting rainwater to tending a garden — foster a deeper connection with the natural world. This relationship encourages:

- **Reduced Carbon Footprint:** By relying on renewable energy sources and sustainable practices, off-grid living directly reduces the strain on our planet's resources.

- **Waste Minimization:** Off-grid homes often adopt composting, recycling, and other waste-reducing strategies, which limits their environmental impact.

- **Biodiversity Boost:** Many off-gridders cultivate native plants, create habitats for local wildlife, and avoid pesticides, enhancing the biodiversity of their surroundings.

A Return to Simplicity

Modern life, with its endless barrage of notifications, commitments, and stresses, can be overwhelming. Off-grid living, in contrast, offers:

- **Mental Clarity:** The simplicity of the lifestyle allows for a decluttered mind. With fewer distractions and stresses, many find it easier to focus on what truly matters.

- **Physical Well-being:** Growing your own food, fetching water, and other daily chores are not only fulfilling but also physically invigorating. The lifestyle naturally promotes physical activity and healthier eating.

- **Emotional Resilience:** Overcoming challenges, from mending a broken solar panel to preserving food for the winter, instills a sense of accomplishment and resilience.

Financial Freedom

Off-grid living can also be financially liberating:

- **Reduced Utility Bills:** Without monthly electricity, water, or gas bills, off-gridders often find their living expenses significantly reduced.

- **Self-sustainability:** Growing your own food and generating your own power can greatly reduce dependency on external markets, offering a cushion against economic downturns.

Cultivating Community and Relationships

Far from the misconception of off-grid living as isolating, it often leads to stronger community bonds:

- **Shared Knowledge:** Off-grid communities frequently come together to exchange skills and knowledge, from sustainable farming techniques to renewable energy solutions.

- **Mutual Support:** In areas where off-grid living is common, communities often form mutual aid groups, helping each other during challenging times.

A Life of Mindfulness

Living off the grid fosters mindfulness. Every drop of water used, every kilowatt-hour generated, and every vegetable harvested is a conscious act. This lifestyle naturally encourages:

- **Gratitude:** A deeper appreciation for life's simple pleasures and the planet's bounties.

- **Mindful Consumption:** When resources are personally generated or cultivated, wastefulness becomes less likely.

- **In-the-Moment Living:** The rhythms of nature, from sunrise to sunset, guide daily activities, making life more attuned to the present moment.

CHAPTER 2: Essential Off-grid Systems

Water Harvesting and Purification

Water is foundational to life, not just for hydration but for every aspect of our daily routine, from cooking and cleaning to nurturing gardens and livestock. For those living off the grid, understanding the importance of sourcing and maintaining a clean water supply is paramount. With traditional utilities out of reach, off-grid residents turn to the ancient and innovative practice of water harvesting and subsequent purification.

Harvesting Water: Tapping into Nature's Reservoirs

Rainwater Collection: Rainwater is among the purest sources of water, mainly because it hasn't yet traveled through ground and contaminants. Off-grid homes often employ roof-based catchment systems where rain is directed into gutters and then stored in cisterns or barrels. Roof materials and design can significantly influence the quality and quantity of rainwater harvested. Safe materials and a large surface area can help ensure a consistent and uncontaminated supply.

Groundwater and Wells: In areas with accessible groundwater, wells become a vital resource. Drilling a well requires understanding the land's topography and aquifer locations. The water extracted typically comes from water-bearing rocks or substrates called aquifers, which act as nature's filters, providing relatively cleaner water than surface sources.

Surface Water Collection: Streams, rivers, and lakes can be essential water sources, especially for larger off-grid communities. When relying on surface water, it's essential to consider its origin and potential contaminant sources upstream. Often, this water requires more rigorous purification processes than rain or groundwater.

Purifying the Essence: Ensuring Water's Safety

Filtration Systems: The primary step in purifying water is filtration. Using various mediums, like sand or activated charcoal, filters work by trapping contaminants as water flows through them. The effectiveness of a filter is often determined by its pore size, with smaller pores being more efficient at catching microscopic pathogens.

Boiling: An age-old method, boiling water is a reliable way to eliminate most pathogens. It's a time-tested technique, but it does consume energy, whether from wood, gas, or solar-heated systems.

Distillation: Mimicking nature's water cycle, distillation involves boiling water and then cooling the resultant steam back into liquid, leaving contaminants behind. It's highly effective, especially in eliminating salts and heavy metals, but requires a substantial energy input.

Ultraviolet (UV) Treatment: Modern off-grid homes may incorporate UV treatments. When water is exposed to UV light, it destroys the DNA of microorganisms, rendering them harmless. This method is energy-efficient and highly effective, but it's best combined with filtration to eliminate non-biological contaminants.

Conserving and Reusing: A Holistic Approach to Water

Given the effort and resources involved in harvesting and purifying water, conservation becomes integral to off-grid living. Practices such as using low-flow fixtures, aerators, and gray water systems not only reduce water consumption but also lessen the strain on purification systems. Furthermore, being mindful of water usage, fixing leaks promptly, and adopting water-efficient gardening techniques, like xeriscaping or drip irrigation, can further reduce water demands.

Waste Management

Waste management in the modern world often remains out of sight and out of mind for many, thanks to efficient municipal systems. However, for those seeking independence from urban grids, managing waste becomes a direct responsibility. Adopting a proactive approach towards waste not only safeguards the immediate environment but also upholds the ideals of sustainable living.

The Off-grid Waste Landscape: Broadening Perspectives

Waste, in an off-grid setting, isn't just about what we typically toss in the garbage. It encompasses everything from kitchen scraps and sewage to packaging materials and old electronics. Unlike urban settings with established infrastructure, off-grid environments require a holistic approach where waste is seen as a potential resource, rather than merely a disposal problem.

Organic Waste: Turning Scraps into Gold

Composting: One of the most common and effective ways to manage organic waste is through composting. By allowing organic materials to decompose naturally, composting turns kitchen scraps, yard waste, and more into nutrient-rich humus that can replenish garden soils. Beyond traditional composting bins, techniques like vermicomposting, using worms, can further accelerate decomposition and enhance nutrient profiles.

Anaerobic Digestion: For those with a bit more technical inclination, anaerobic digesters provide a dual solution. In the absence of oxygen, organic waste breaks down, producing biogas, which can be harnessed for cooking or heating, while the residue, termed digestate, acts as a potent organic fertilizer.

Solid Waste: Navigating Non-Biodegradables

Reducing Consumption: Prevention is always better than cure. By making conscious purchasing choices and opting for products with minimal or eco-friendly packaging, the generation of solid waste can be drastically reduced.

Repurposing and Upcycling: Before discarding, considering if an item can be given a second life is essential. Old containers can become storage solutions, while worn-out textiles might be transformed into quilts or rugs.

Landfills, Burying, and Burning: While less ideal, in certain off-grid scenarios, creating a small controlled landfill might be necessary. Burying waste is another method, but care should be taken to avoid contaminating water sources. As for burning, it's vital to understand that not all waste is suitable for incineration due to the release of toxic fumes. Only burn when necessary and ensure proper ventilation.

Liquid Waste and Sanitation: Vitality in Cleanliness

Constructed Wetlands: Mimicking natural processes, constructed wetlands allow wastewater to pass through a series of aquatic plants that naturally filter and break down contaminants.

Composting Toilets: A solution to both liquid and solid human waste, composting toilets transform excrement into compost over time, eliminating pathogens and recycling nutrients.

Redefining Waste: A Philosophical Shift

For the off-grid enthusiast, waste isn't a by-product but rather a challenge to reimagine what's possible. By viewing waste as a potential resource, the narrative shifts from disposal to cyclical reuse. Whether it's transforming kitchen waste into garden gold or repurposing old jars into lanterns, off-grid waste management is as much about creativity as it is about responsibility.

CHAPTER 3: Building an Off-grid Home

Energy Efficient Design

An energy efficient design ensures that the limited resources available are utilized to their utmost potential, creating a dwelling that harmoniously melds with nature, rather than drawing incessantly from it.

The Art of Passive Solar Design

Sunlight, a bounteous and ever-reliable gift from nature, plays a pivotal role in the design of energy efficient homes. Passive solar design, a strategy that involves orienting a building to capture, store, and distribute solar energy, optimally heats a home during cold months and deflects excessive heat during the warmer seasons. By positioning windows, walls, and floors to collect and store solar energy and then distribute the heat generated, homes can naturally regulate temperature, reducing the need for additional heating or cooling systems. Overhanging roofs, when appropriately placed, can also ensure that high summer sun is blocked, while the lower winter sun is allowed to penetrate and warm living spaces.

Insulation: The Unsung Hero

A home's ability to maintain its internal temperature is significantly influenced by its insulation. Properly insulated walls, roofs, and floors can prevent warm air from escaping during chilly nights and barricade against heat during sweltering days. Materials like straw bales, cellulose, and even sheep's wool, offer natural insulation solutions, ensuring a cozy interior while keeping the ecological footprint minimal.

Harnessing the Earth: The Potential of Thermal Mass

Materials like brick, stone, and concrete have an inherent ability to absorb, store, and release heat over time, a quality known as thermal mass. When used strategically in flooring or walls, these materials can soak up heat during the day and gradually release it at night, acting as natural temperature regulators.

This principle, deeply rooted in ancient architectural practices, finds renewed relevance in energy efficient off-grid homes.

The Breath of the Home: Ventilation and Airflow

A well-designed home does more than merely protect its inhabitants from the elements; it breathes. Incorporating features such as clerestory windows, skylights, and strategically positioned vents can promote natural airflow, facilitating the circulation of fresh air and aiding in temperature regulation. Such designs, besides ensuring energy efficiency, also contribute to healthier indoor air quality.

A Thoughtful Approach to Lighting

Natural light, with its soft glow and mood-enhancing properties, is an invaluable asset in energy efficient design. By integrating larger windows, skylights, and light-reflecting surfaces, the dependency on artificial lighting during the day can be markedly reduced. For nighttime illumination, LED lights, which consume a fraction of the energy of traditional bulbs, can be incorporated, ensuring that light is available without a hefty energy price tag.

Water Efficiency: Beyond Just Energy

While the focus has largely been on energy in terms of heating, cooling, and lighting, water too plays a role in the efficiency narrative. Harvesting rainwater, employing low-flow fixtures, and reusing greywater can drastically reduce water consumption, ensuring that this precious resource is judiciously used and conserved.

Incorporating Renewable Energies

Drawing power from nature, be it the sun, wind, or water, allows homeowners to craft a self-sufficient sanctuary, a place where they are not tethered to conventional energy grids. Instead, they're powered by the ceaseless rhythms of the natural world. Here, we delve into how one can seamlessly weave renewable energy systems into the fabric of an off-grid home.

Solar: The Powerhouse of the Off-grid Domain

Solar energy, with its ubiquity and ever-increasing efficiency, is often the first choice for off-grid homes. Integrating solar panels, whether on rooftops, walls, or ground mounts, provides a dependable source of electricity. But it isn't just about placing panels; it's about understanding the sun's trajectory, maximizing sun hours, and employing efficient energy storage solutions to ensure a consistent power supply, even when the skies are overcast.

Wind Energy: Harnessing the Breezes

While solar energy is a go-to for many, wind energy can also play a complementary or primary role, depending on the location. Wind turbines, ranging from towering structures to more compact designs suitable for residential use, can be integrated into properties that experience consistent wind patterns. The gentle (or sometimes robust) gusts of wind can be transformed into electrical energy, filling the home with the whisper of nature's power.

Hydro: The Grace of Flowing Water

For those fortunate enough to have flowing water on their properties, micro-hydro power systems can be a game-changer. These systems, which harness the kinetic energy of flowing water to generate electricity, offer a consistent power source, as water flows irrespective of day or night, sun or cloud. The gentle hum of a turbine, powered by a nearby stream or river, can provide not only energy but also a sense of profound connection to the land.

Geothermal: Delving Deep into the Earth's Embrace

Beneath the surface of our planet lies a consistent source of heat. Geothermal energy systems, which harness this heat, can be integrated into off-grid homes, providing both heating during colder months and cooling during warmer seasons. By circulating a fluid through pipes buried deep underground and then extracting the heat from this fluid, homes can enjoy temperature regulation powered by the very heartbeat of the Earth.

Biomass: Rekindling Ancient Practices

Drawing inspiration from time-honored practices, biomass energy solutions involve burning organic materials, such as wood, agricultural residues, or even animal dung, to produce heat. This heat can be used for cooking, heating homes, or even generating electricity in more sophisticated setups. While it's vital to source materials sustainably and ensure efficient combustion to minimize emissions, biomass can offer a touch of primal energy to the modern off-grid home.

Optimizing Energy Consumption: The Dance of Demand and Supply

Incorporating renewable energy systems is only one side of the coin. Equally crucial is understanding and optimizing energy consumption within the home. Energy-efficient appliances, thoughtful usage patterns, and monitoring systems can ensure that the energy generated is used judiciously, creating a harmonious balance between what the Earth provides and what the inhabitants consume.

CHAPTER 4: Off-grid Communities

Worldwide Off-grid Movements

Across continents, cultures, and communities, a profound transformation is underway. There's a growing movement of individuals and groups choosing to detach from traditional energy grids and embrace self-sustained living. This worldwide off-grid movement isn't just about generating one's electricity; it's about reshaping our relationship with resources, fostering communities, and nurturing a deeper connection to the natural world.

A Glimpse into Various Corners of the Globe

Europe: Off-grid Villages and Eco-communities
In countries like Spain, Portugal, and the UK, off-grid living has experienced a resurgence, particularly in rural and remote regions. People are forming eco-communities, where they share resources, skills, and the commitment to a reduced carbon footprint. These communities often incorporate permaculture principles, ensuring that the land is nurtured, and resources are utilized sustainably.

Africa: Solar-Powered Dreams and Grassroots Initiatives
Given its abundant sunshine, Africa has seen a significant rise in solar-powered off-grid solutions, especially in regions where conventional electricity has been elusive. Grassroots initiatives, often supported by international organizations, are lighting up homes, powering schools, and transforming communities. It's a tale of empowerment, where people, once in darkness, now bask in the brilliance of solar innovation.

Asia: Balancing Tradition with Modern Off-grid Solutions
In Asia, off-grid living takes on varied hues. From the pristine Himalayan villages harnessing hydropower to the remote islands in the Philippines utilizing wind and

solar, there's a confluence of traditional wisdom and modern innovation. Many Asian communities, drawing from centuries-old practices, seamlessly integrate renewables, ensuring that their cultural heritage thrives alongside technological advancements.

The Americas: Pioneering Off-grid Lifestyles and Eco-villages

Both North and South America present a tapestry of off-grid lifestyles. From individuals pioneering self-sufficient lives in the expansive terrains of Canada and the USA to eco-villages sprouting in the lush landscapes of Costa Rica and Brazil, there's a shared spirit of independence. Many are drawn to these lifestyles for environmental reasons, while others seek a deeper community connection or simply the freedom of self-reliance.

Challenges, Triumphs, and the Shared Spirit of Resilience

While the romantic notion of off-grid living appeals to many, it's essential to recognize that it comes with its share of challenges. From initial setup costs to ensuring consistent energy supply and facing extreme weather conditions, off-gridders often have to be innovative and resilient.

However, the triumphs often outweigh the tribulations. There's the joy of crafting a life on one's terms, the fulfillment of nurturing the environment, and the profound community bonds forged in shared aspirations and mutual support.

Moreover, as technology advances, many of the challenges associated with off-grid living are becoming more manageable. Innovations in energy storage, efficient appliances, and community-driven solutions are ensuring that more people can embrace this lifestyle with confidence.

Building Community Resilience

Resilience is often depicted as the ability to rebound from adversities, but at the core of this attribute lies a deeper essence, especially when referred to in a community context. Community resilience embodies the combined strength, adaptability, and innovation of its members, making it capable of withstanding and evolving through challenges, be they environmental, economic, or societal.

The Imperative of Strengthened Communities in Off-grid Living

When it comes to off-grid living, the importance of a resilient community becomes even more paramount. Given the nature of this lifestyle, with its reliance on localized resources and its detachment from broader infrastructural supports, communities become the backbone of sustenance and progress. The challenges unique to off-grid living, such as resource limitations, extreme weather events, or even the isolation of remote locations, require a collective response that is agile, cohesive, and forward-looking.

Cultivating Community Bonds: The Role of Shared Values and Vision

A resilient off-grid community often finds its strength in shared values and a common vision. When individuals come together, motivated by a mutual respect for nature, a commitment to sustainable practices, and a desire for genuine interdependence, they lay the foundation for a community that can endure hardships and thrive in evolving circumstances.

Moreover, the process of building and maintaining an off-grid lifestyle necessitates collaboration. From setting up renewable energy systems to managing water resources, from crafting communal spaces to organizing events that foster bonding, every aspect becomes an opportunity for community members to learn from one another, share their expertise, and build deeper connections.

Empowerment through Education and Skill Sharing

One of the cornerstones of community resilience in off-grid settings is the continuous process of education and skill sharing. Given that off-grid living can demand a wide range of skills – from understanding solar panel installations to organic farming techniques – communities benefit immensely from internal workshops, mentorship programs, and knowledge exchange sessions. When every member feels empowered with knowledge and skills, not only does the community's collective capability grow, but its resilience is also fortified.

Economic Resilience: Local Economies and Barter Systems

While the essence of community resilience is often rooted in shared values and social bonds, economic aspects cannot be overlooked. Many off-grid communities around the world have innovated local economic models that prioritize sustainability and mutual benefit over profit. Concepts like community-supported agriculture, local currencies, and barter systems come to the fore, ensuring that the economic foundations of these communities are as resilient as their social ones.

Facing the Future: The Role of Forward Planning and Preparedness

No matter how cohesive and skilled a community might be, the unforeseen challenges of the future demand a level of preparedness. Off-grid communities, given their inherent vulnerabilities to factors like climate change, often prioritize forward planning. Whether it's creating contingency plans for extreme weather events, building buffer stocks of essential resources, or devising strategies for sustainable growth, forward planning becomes an integral part of community resilience.

BOOK 7

Urban Solar Solutions

CHAPTER 1: Solar in the City

Challenges and Opportunities

The adoption of solar energy in urban areas presents an intricate dance between the promises of renewable energy and the complexities inherent in city structures. As cities worldwide seek to reduce their carbon footprint and drive towards sustainability, solar energy emerges as an indispensable ally. However, harnessing the sun's power amidst towering skyscrapers, congested streets, and the frenzied pace of city life comes with its own set of challenges and opportunities.

Challenges in the Urban Solar Quest

Space Constraints: One of the most apparent challenges is the limited availability of space. Unlike rural or suburban settings where open lands or expansive rooftops might be available, cities often deal with cramped spaces, making it tricky to install large solar arrays.

Shadowing and Building Heights: The varying heights of buildings mean that a solar installation on one building might be in the shadow of a taller neighboring structure, significantly reducing its efficiency. The dynamic nature of city structures, with new buildings cropping up, further complicates the task of ensuring consistent solar exposure.

Regulations and Policies: Cities often have a complex web of regulations governing construction and modifications to existing structures. Incorporating solar installations requires navigating these rules, some of which might not have been updated to accommodate the nuances of solar energy.

Initial Costs: Urban environments, with their high real estate prices and installation complexities, can sometimes drive up the initial costs of solar projects, posing challenges for both individuals and businesses contemplating the switch.

Yet, Every Cloud Has a Silver Lining: The Opportunities

High Energy Demand: Cities, being hubs of commercial, residential, and industrial activities, have significant energy demands. This high demand means that even a small percentage shift to solar can result in substantial absolute energy generation from renewable sources.

Innovative Installations: The space constraints of urban areas have spurred innovation. From solar windows to vertical installations on building facades, cities have witnessed some of the most creative solar solutions, turning challenges into opportunities.

Policy Incentives: Recognizing the need for a sustainable future, many city administrations offer incentives, grants, and favorable policies to promote solar adoption. Such measures can offset the initial costs and make solar projects more financially viable.

Community Engagement: Urban areas, with their dense populations, provide an excellent platform for community solar projects. Shared solar initiatives, where multiple households or businesses invest in a joint solar project, can pool resources, share costs, and enjoy the benefits of solar energy without individual installations.

Green Branding: For businesses in urban areas, adopting solar is not just an energy decision but also a branding one. With consumers increasingly favoring eco-friendly brands, businesses can leverage their solar initiatives as a testament to their commitment to sustainability.

The Path Forward: Balancing Act in the Urban Canopy

Urban solar adoption is neither a straightforward challenge nor a mere opportunity – it's a blend of both. While the road to extensive solar adoption in cities might be riddled with obstacles, the potential benefits – both environmental and economic – make the journey worthwhile.

Urban Solar Innovations

Urban areas, with their intrinsic challenges, have often been the breeding ground for ingenuity and innovation. The realm of solar energy is no exception. As the urgency for sustainable energy solutions intensifies, urban centers worldwide are introducing novel techniques and groundbreaking technologies to integrate solar power seamlessly into the cityscape.

Integrated Photovoltaics: Not Just a Window to the World

Integrated photovoltaics (BIPV) have emerged as a game-changer. BIPV are solar cells incorporated into construction materials, most notably windows and facades. These are not mere add-ons; they are integral parts of the building. Transparent solar panels can replace traditional windows, allowing buildings to generate electricity without compromising on aesthetics. As more skyscrapers adopt BIPV, cities might soon see their skyline transformed into vertical power stations.

Solar Roads and Pathways: Walking on Sunshine

While rooftops and facades are the most obvious spots for solar panels, innovators are looking downward—to the streets. Some cities have started testing solar roads and pathways. These are made of durable solar panels covered with a protective surface sturdy enough to withstand the weight of vehicles. As cars traverse these roads, underneath, the sun's energy is captured and converted to electricity. Given the vast stretches of roads in urban areas, this innovation has the potential to contribute substantially to a city's energy needs.

Solar-powered Street Furniture: Functional and Sustainable

Bus stops, benches, and street lamps have undergone a solar makeover in several cities. These everyday urban elements are now embedded with solar panels, serving dual purposes: providing essential services and generating renewable energy. A solar bench, for instance, might allow you to charge your phone while you rest, using the power it harnessed earlier in the day.

Floating Solar Farms: Making Waves in Energy Generation

With land space at a premium, some cities are turning to their water bodies. Floating solar farms, installed on lakes or reservoirs within city limits, are an ingenious solution to space constraints. Besides generating electricity, they also reduce water evaporation and inhibit the growth of harmful algae by blocking sunlight.

Solar-powered Public Transport: Riding the Sun's Rays

Public transportation, the lifeline of any major city, is also joining the solar revolution. We're witnessing the advent of solar-powered buses and trams. Not only are their roofs lined with solar panels, but many transport hubs, like bus depots and train stations, are also being equipped with solar installations, ensuring a sustainable transit system.

Urban Solar Canopies and Overpasses: Shelter and Power

Parking lots and pedestrian overpasses are often seen as dead spaces in terms of functionality beyond their primary use. However, with the introduction of solar canopies, these spaces are being transformed into power generation centers. These canopies, apart from generating power, provide shade and shelter, enhancing urban functionality.

CHAPTER 2: Vertical Solar Installations

Solar Windows and Facades

In the heart of sprawling urban landscapes, where horizontal space is a precious commodity, vertical surfaces present a wealth of untapped potential. The vertical integration of solar technology, especially in the form of solar windows and facades, offers a promise of revolutionizing urban energy solutions, merging aesthetics with functionality.

The Genesis of Solar Windows and Facades

The concept of solar windows and facades isn't entirely novel. Historically, the idea of harvesting solar energy from vertical surfaces has been an aspiration, but technological limitations restrained its fruition. However, with advancements in material science and photovoltaic technologies, the once distant dream is gradually materializing into tangible reality.

A Transparent Transition

Solar windows, at their core, represent the amalgamation of form and function. These aren't your standard opaque solar panels. Instead, they are transparent or semi-transparent photovoltaic glass panels that replace conventional windows. This transformative technology ensures that while buildings receive sunlight for illumination, they also covert part of the sun's rays into electricity.

Organic photovoltaics, a subset of solar technology, plays a pivotal role in this. By using organic materials (like polymers) instead of traditional silicon, these panels can be made semi-transparent, lightweight, and versatile. The thin-film technology incorporated allows for flexibility, meaning these solar panels can be integrated seamlessly into various architectural designs without compromising aesthetic appeal.

Facades: The Vertical Powerhouses

While solar windows focus on blending transparency with power generation, solar facades are more about maximizing energy output. Covering significant portions of a building's exterior, facades are ideally positioned to capture sunlight throughout the day. Given their expansive surface area, even with a slightly lower efficiency compared to conventional rooftop panels, their cumulative energy generation can be substantial.

Most solar facades employ thin-film solar cells or crystalline silicon cells, depending on the design requirements. The former offers more flexibility and can be incorporated into curved or uniquely shaped structures, while the latter provides a higher energy yield. The choice often boils down to the architectural intent: is it more about aesthetic integration or maximizing power generation?

While the advantages of solar windows and facades are many, challenges persist. The efficiency of transparent photovoltaics, though improving, still lags behind traditional solar panels. Costs, too, remain relatively high, though economies of scale and further research promise reductions in the future.

Yet, the potential impact of these vertical solar solutions on urban energy dynamics is undeniable. As more architects, builders, and city planners recognize the multifaceted benefits – from energy savings and reduced carbon footprints to enhanced building aesthetics – the urban skyline of the future might just shimmer with the promise of sustainability.

Solar Balconies and Awnings

In such spaces, every inch is prime real estate. Balconies, traditionally seen as spaces for relaxation and a touch of the outdoors, are now emerging as the new frontier in the realm of solar solutions. Similarly, awnings, which have been used for centuries to provide shade, are getting a modern makeover as solar harvesters.

Balconies: The Elevated Solar Gardens

A balcony can be much more than just a space to sip your morning coffee. With the right technology, it can transform into a mini power plant. The idea is simple: use the balcony's railings and floors to mount thin, flexible solar panels. These panels can capture sunlight and convert it into electricity for immediate use or storage.

With solar balconies, residents have the power to generate a portion of their daily electricity needs, decreasing their reliance on the grid. For apartments where rooftop access is often restricted, or where the roof is already crowded with various amenities, balconies offer an accessible and personal space for solar installation.

However, it's not just about laying down solar panels haphazardly. Innovative designs are emerging which incorporate solar panels into balcony railings, flooring, and even the walls, ensuring maximum sunlight capture while preserving the aesthetic appeal of the space.

Awnings: Dual-Purpose Shades

While awnings have always served the primary function of providing shade and reducing heat, integrating solar technology transforms them from passive structures to active energy generators. Solar awnings are essentially retractable awnings fitted with lightweight, flexible solar panels. Positioned optimally, they can capture a significant amount of sunlight.

An added advantage of solar awnings is their adaptability. They can be extended or retracted based on the sun's position and the desired amount of shade, ensuring a dynamic approach to energy capture. This not only optimizes energy generation but also provides a customizable shade solution for urban homes and businesses.

Embracing the Blend of Function and Design

The challenge with both solar balconies and awnings isn't just technological – it's also about design integration. The goal is to seamlessly blend solar technology with architectural and design elements, ensuring that these installations enhance the visual appeal of urban structures. Thankfully, with advancements in solar panel flexibility and design variety, there's ample room for creativity.

Moreover, the introduction of smart systems can amplify the benefits of these installations. Imagine a smart solar awning that adjusts its angle based on the sun's position, or a balcony system that provides real-time feedback on energy generation – the possibilities are vast and exciting.

CHAPTER 3: Community Solar Projects

Shared Solar Benefits

City landscapes are often characterized by closely clustered buildings, bustling streets, and the vibrant energy of community life. While this closeness has led to shared experiences, memories, and events, the concept of sharing has now taken a leap into the domain of energy as well. Shared solar, also commonly known as community solar, is an evolving movement that seeks to redefine how urban populations interact with, consume, and benefit from solar energy.

A Collective Approach to Solar Energy

Shared solar projects are a collective solution to a common urban problem: limited individual access to sunlight due to high-rises, shaded areas, or lack of individual spaces suitable for solar installations. Instead of each apartment owner or renter struggling with individual solar setups, a shared system is established, often on a communal piece of land, a spacious rooftop, or an open field nearby. The electricity generated from this communal solar array is then distributed among the participating members, either offsetting their utility bills or providing direct power.

Unlocking a Myriad of Benefits

One of the primary advantages of shared solar projects is that they make solar energy accessible to a wider demographic. Not everyone owns a house with a sun-facing roof. Renters, apartment dwellers, and even homeowners with shaded roofs can tap into the benefits of solar energy without having to install personal systems. Financially too, shared solar systems can be more viable for many. By pooling resources, community members can share the initial setup costs, thereby reducing the individual financial burden. Over time, the returns on this shared investment can be significant, especially as utility prices rise.

Furthermore, maintenance and upkeep become streamlined in a shared system. Instead of multiple individual setups that need regular checks and servicing, a singular, larger system can be monitored and maintained more efficiently.

From an environmental perspective, large-scale community solar projects can lead to a significant reduction in carbon footprints. As more urban communities gravitate towards such shared solutions, cities can witness a sizable shift towards clean energy, thus aligning with broader goals of sustainability and reduced greenhouse gas emissions.

Strengthening Community Bonds

Beyond the tangible benefits of energy and cost savings, shared solar projects foster a sense of community. Collaboratively working towards a shared goal, managing the system, and reaping collective benefits can strengthen community bonds. Decision-making becomes a collective process, fostering discussions, collaborations, and a sense of shared responsibility.

Additionally, these projects can serve as educational hubs. As the community gets involved, especially the younger generation, there's a growing understanding of renewable energy, sustainability, and the intricacies of energy management.

A Blueprint for a Brighter Urban Future

As cities continue to grow and evolve, so do their energy needs. Traditional energy solutions, often reliant on fossil fuels and centralized grids, come with a plethora of challenges, from pollution to price surges. Shared solar projects, in contrast, paint a promising picture of what urban energy dynamics can look like.

With the dual advantages of environmental sustainability and financial viability, coupled with the potential for community-building, shared solar stands out as a beacon of hope in the urban energy landscape. It's a testament to the power of collective action and the endless possibilities that emerge when communities come together for a brighter, cleaner future.

Setting Up a Community Solar Project

Community solar projects are rapidly gaining traction, offering urban dwellers an alternative way to harness the sun's energy. But like any major project, setting up a community solar initiative requires careful planning, coordination, and an understanding of the various stages involved.

Initiation: Assessing Interest and Building a Team

The first step in setting up a community solar project is gauging interest within the community. Organize town hall meetings, circulate surveys, or hold informational sessions to understand how many members of the community are willing to participate.

Once there's a clear interest, assemble a dedicated team or committee to oversee the project. This team should ideally consist of individuals with diverse skills, from project management to technical expertise in solar energy.

Feasibility Study: Determining Project Viability

Before jumping into the project, conduct a feasibility study. This should include:

- **Location Analysis:** Identify potential sites for the solar installation. This could be unused community land, a large common rooftop, or other suitable spaces. Consider factors like sun exposure, obstructions, and proximity to the community.

- **Cost Estimation:** Determine the overall budget for the project. This includes the cost of solar panels, inverters, mounting structures, and other necessary equipment. Don't forget to factor in the costs of installation, maintenance, and potential upgrades.

- **Energy Yield Predictions:** Estimate the potential energy that the solar installation could generate. This can be done using solar mapping tools or by consulting with solar energy experts.

Engaging Stakeholders: Building Community Buy-in

It's essential to keep the community involved and informed throughout the process. Regularly update them on the project's progress, the expected benefits, and how the costs and benefits will be shared among the participants.

Engaging local authorities or city councils can also be beneficial. They may offer incentives, grants, or other forms of support for community-driven renewable energy projects.

Choosing a Business Model: Financial Frameworks

There are multiple ways a community solar project can be financially structured:

- **Ownership Model:** Community members collectively invest and own the solar installation. The benefits (electricity or monetary savings) are then distributed based on individual investments.

- **Subscription Model:** Instead of owning the infrastructure, community members can subscribe to the solar service, typically through monthly payments. They then receive benefits based on their subscription level.

- **Third-party Ownership:** An external entity invests in and owns the solar setup, while the community enters into an agreement to purchase the generated power, often at reduced rates.

Implementation: Bringing the Vision to Life

With planning in place, it's time for implementation. This phase involves:

- **Vendor Selection:** Choose a reliable solar service provider. Look for companies with experience in community solar projects and a track record of quality installations.

- **Permitting and Compliance:** Ensure that the project adheres to local regulations. This might involve obtaining necessary permits or passing specific inspections.

- **Installation:** Once all approvals are in place, the solar service provider will begin the installation process. This will include setting up solar panels, inverters, and connecting the system to the grid or storage solutions if necessary.

Ongoing Management and Maintenance

Once the system is operational, it's crucial to monitor its performance and maintain it. The dedicated team or committee should oversee this, ensuring that the system operates at peak efficiency. Regular maintenance checks, cleaning of panels, and addressing any technical issues are vital.

Cultivating a Lasting Legacy

Setting up a community solar project is more than just harnessing clean energy; it's about creating a lasting legacy of sustainability. Through the shared commitment of a community, these projects can set a precedent, inspire other communities, and contribute significantly to a greener urban future.

CHAPTER 4: Urban Farming and Solar

Green Roofs and Solar Integration

The pressing need for sustainable urban solutions has given rise to innovative ideas, and among these, the integration of green roofs with solar panels stands out. This synthesis promises both ecological balance and energy efficiency, tailoring a response to urban heat islands, and the energy crisis.

Understanding Green Roofs

Green roofs, often termed 'living roofs,' involve growing vegetation on building rooftops. While traditionally associated with aesthetics or recreational spaces, these roofs have proven their worth far beyond mere ornamental value. They help in temperature regulation, improving air quality, and managing stormwater runoff. In essence, green roofs act as lungs for cities, breathing in carbon dioxide and exhaling fresh oxygen.

The Complementary Nature of Solar Panels

Solar panels, on the other hand, have been a beacon of renewable energy for decades. Their presence in urban areas is indispensable, considering the pressing need to shift from fossil fuels. When installed on rooftops, they capture sunlight and convert it into electricity, powering homes and reducing reliance on the grid.

Melding the Two: The Symbiotic Relationship

At first glance, it might seem counterintuitive to pair lush greenery with sleek solar panels, but there's a harmonious relationship to be uncovered:

- **Temperature Regulation:** One of the challenges solar panels face in urban settings is overheating, which can reduce their efficiency. The cool microclimate created by green roofs can help regulate the temperature, ensuring that solar panels perform optimally.

- **Maximizing Roof Usage:** Not all parts of a roof receive equal sunlight. Those shaded areas, unsuitable for solar panels, can be utilized for vegetation. This ensures that every inch of the roof is used productively.

- **Enhanced Aesthetics:** The blend of technology and nature offers a visual treat. Urban dwellers often yearn for green spaces, and a roof that offers both verdant beauty and technological prowess is a welcome sight.

- **Biodiversity and Energy:** While the vegetation attracts and supports local biodiversity, the solar panels harness clean energy. Together, they contribute to a sustainable and life-affirming urban environment.

Challenges in Integration

While the combination promises numerous benefits, there are hurdles to overcome. Weight load is a primary concern, as roofs must support both the soil, vegetation, and solar installations. Waterproofing is another crucial aspect, ensuring that neither the vegetation nor the solar setup harms the building's integrity.

Moreover, maintenance becomes a dual task. Both the plants and the solar panels need regular care, which requires an integrated approach to ensure neither is compromised.

Future Implications for Urban Planning

The integration of green roofs with solar panels represents a bold step forward in urban planning. As cities grapple with space constraints, rising temperatures, and energy needs, solutions like these address multiple challenges simultaneously.

City planners, architects, and sustainability experts are taking note of these integrated setups. They see potential not just in terms of energy generation and ecological balance, but also in fostering community engagement. Rooftops can become communal spaces, where residents gather, children learn about nature and renewable energy, and communities bond over shared values.

Urban Solar Greenhouses

The quest for sustainable food production and energy consumption has taken an innovative turn with the emergence of urban solar greenhouses. As urban sprawls tighten their grip, relegating agricultural lands to the peripheries, these greenhouses present a harmonious blend of solar energy harnessing and localized food production. It's the dream merger of technology and agriculture, tailored for the concrete jungles we inhabit.

Understanding Urban Greenhouses

Traditional greenhouses have been pivotal in food production, allowing for controlled environments where plants can thrive irrespective of external weather conditions. In urban settings, these greenhouses have taken on vertical designs, optimizing space in a bid to produce more in confined areas. They protect crops from pollutants, pests, and erratic weather, ensuring consistent yields and reducing the need for chemical interventions.

Solar Integration: A Game Changer

Solar technology's inclusion into urban greenhouses amplifies their functionality. Rather than just being spaces for food production, they morph into power generators, achieving a dual objective. Here's how:

- **Translucent Solar Panels:** Innovations in solar technology have birthed translucent solar panels that can be integrated into greenhouse roofs. While these panels harvest sunlight for energy, they still allow a percentage of light to penetrate, essential for plant growth.

- **Energy Autonomy:** Urban greenhouses, especially those running hydroponic or aeroponic systems, require energy for water pumps, temperature control, and lighting. By harnessing solar energy, these greenhouses can potentially become energy-autonomous, reducing operational costs and carbon footprints.

- **Optimal Light Wavelengths:** Solar panels can be tailored to filter and allow only specific light wavelengths that plants predominantly use for photosynthesis, ensuring efficient plant growth while maximizing energy capture.

The Multifaceted Benefits

The integration of solar technology into urban greenhouses isn't just a marriage of convenience—it's one of necessity and foresight. Here's what this synergy promises:

- **Local Food Production:** As transport becomes a major contributor to carbon emissions, producing food closer to consumers drastically cuts down the carbon footprint. It also ensures fresher produce, devoid of long storage durations and transportation lags.

- **Reduced Grid Dependence:** As cities grapple with increasing energy demands, every kilowatt-hour produced off the grid counts. Solar greenhouses contribute to this off-grid energy, relieving pressure on city power infrastructures.

- **Education and Awareness:** Urban solar greenhouses can be centers of learning. They provide urban dwellers, especially the younger generation, with insights into food production, renewable energy, and sustainability. It's an experiential learning hub right in the heart of the city.

- **Economic Opportunities:** These greenhouses can spur job creation, offering roles in both the agricultural and technological sectors. They can become hubs for research, innovation, and commerce, selling produce directly to consumers.

While promising, the integration isn't without challenges. The initial investment required for such greenhouses can be substantial. Balancing light requirements for both plants and solar energy capture is a nuanced task. Maintenance, especially in polluted urban areas, can be demanding.

Smart Farm

Green Industry

Green City

BOOK 8

Mobile Solar Setups

CHAPTER 1: Solar on the Move

Portable Solar Panels

The rise of portable solar panels has signaled a monumental shift in our relationship with energy. Not confined to rooftops or vast fields, these panels bring the promise of solar energy into the palm of our hands, ensuring that we're never far from a renewable power source, irrespective of our geographical location.

Understanding Portable Solar Panels

Unlike their larger, more fixed counterparts, portable solar panels are designed for mobility. They're lightweight, foldable, and optimized for individual use. Manufactured from thin layers of photovoltaic cells, these panels can be rolled, folded, or hung, making them versatile companions for various applications.

The Evolution of Portability

From the initial rigid, cumbersome panels that could be moved but not easily, today's portable panels are a testament to technological progression. Scientists and manufacturers have been diligently at work to reduce their weight while boosting their efficiency, ensuring that these panels are both easy to carry and effective in their function.

Materials such as CIGS (Copper Indium Gallium Selenide) have replaced traditional silicon in many portable panels. This has not only reduced their weight but has also allowed for flexibility. The result? Solar sheets that can be folded like a cloth or rolled like a scroll.

Applications and Use Cases

The advent of portable solar panels has broadened the horizons of solar energy application. Some of the prominent uses include:

- **Camping and Hiking:** Adventurers no longer need to rely solely on batteries or campfires. With portable panels, they can charge their devices, cook meals, and even heat water, all while being immersed in nature.

- **Military Operations:** Soldiers on covert operations, especially in remote terrains, can leverage these panels for their energy needs without the logistical nightmares associated with fuel transportation.

- **Journalism and Exploration:** Reporters covering stories in off-grid locations or explorers charting unmarked territories can use these panels to power their equipment.

- **Everyday Urban Use:** Not all applications are in remote terrains. Many urban dwellers use portable panels in parks, beaches, or cafes, ensuring their devices never run out of power.

Benefits and the Promise of Freedom

What truly makes portable solar panels remarkable is the freedom they offer. The freedom to roam without being tethered to power sources, the freedom from reliance on non-renewable energy during travels, and the freedom from electricity bills during short trips.

Moreover, their environmental footprint (or lack thereof) is commendable. By relying on the sun, users can reduce their carbon emissions, making their travels and operations greener.

Challenges and Limitations

Like all technologies, portable solar panels aren't without their challenges. Their efficiency, although improving, is still less than fixed installations. Weather conditions greatly influence their performance, with cloudy or rainy days limiting their energy output.

Moreover, the energy storage solutions (like batteries) that accompany these panels need to be lightweight yet capable, a challenge that manufacturers are still grappling with.

Solar Backpacks and Gadgets

Enter solar backpacks and gadgets — functional, innovative solutions that are redefining the realms of portability and sustainable energy.

The Concept of Solar Backpacks

Imagine walking under the sun and simultaneously charging your devices. That's the promise of solar backpacks. Essentially, they are regular backpacks but come with integrated solar panels on their exterior. These panels absorb sunlight and convert it into electricity, which can then be used to charge various devices.

The Anatomy of a Solar Backpack

The core component of a solar backpack is its photovoltaic cells, which are usually sewn into the fabric of the backpack or are attached as a flexible panel. These cells are connected to a battery, which stores the energy generated. Most solar backpacks come with USB ports, allowing users to charge a wide range of devices, from smartphones and tablets to cameras and GPS units.

Efficiency and Practicality

Solar backpacks are designed for convenience and efficiency. On a sunny day, a typical solar backpack can generate enough electricity to charge a smartphone fully in just a few hours. The stored energy in the battery ensures that devices can be powered even when the sun isn't shining.

However, the charging efficiency can vary based on several factors, including the quality of the solar panels, the intensity of sunlight, and the device's power requirements. A cloudy day or a shaded environment will inevitably slow down the charging process.

Gadgets Going Solar

Beyond backpacks, the portable solar trend has expanded to a myriad of gadgets. Here are some notable inclusions:

- **Solar-powered Chargers:** These are compact, lightweight devices equipped with small solar panels. They can charge a variety of devices and often come with multiple USB ports.

- **Solar-powered Lanterns and Torches:** Especially useful for campers and trekkers, these gadgets ensure you're never in the dark. They charge during the day and provide light at night.

- **Solar-powered Speakers:** Music enthusiasts can now take their tunes everywhere without worrying about batteries. These speakers charge under the sun and can play music for hours.

- **Solar-powered Wearables:** From watches to fitness bands, wearable tech is also embracing solar. These gadgets use ambient light to stay powered, reducing the frequency of charging.

A Sustainable Shift

The proliferation of solar backpacks and gadgets represents a broader shift towards sustainable and renewable energy sources. Users not only get a reliable power source but also contribute to reducing carbon footprints. Every device charged via solar energy is a step away from conventional electricity and its associated environmental impacts.

Challenges and Innovations

The primary challenge with solar gadgets is ensuring consistent energy supply. Sunlight is variable, and not every day is sunny. To combat this, many gadgets are incorporating hybrid charging solutions, allowing users to charge via solar or traditional electricity.

Another challenge lies in aesthetics and design. Solar panels, traditionally, aren't the most visually appealing. Innovations in flexible and transparent solar cells are paving the way for more stylish and integrated designs.

CHAPTER 2: Solar-Powered Vehicles

Solar Cars and Bikes

The challenge of sustainable mobility is front and center in the discourse surrounding global warming and environmental degradation. With the transportation sector being a significant contributor to global carbon emissions, the race is on to develop technologies that can power our journeys in cleaner, more efficient ways. Solar-powered vehicles, especially cars and bikes, have emerged as promising solutions in this pursuit.

The Dawn of Solar-Powered Cars

Solar cars have a somewhat futuristic appeal, but they are no longer relegated to the pages of sci-fi novels or experimental laboratories. These vehicles rely on solar panels, typically integrated into their roofs or bodies, to convert sunlight into electricity. This electricity is then used to power the vehicle's motor, propelling it forward.

How Do Solar Cars Work?

Solar cars primarily use photovoltaic cells to capture sunlight. These cells are made of semiconductor materials, which create an electric field when exposed to sunlight. As photons—light particles—strike these cells, they dislodge electrons, generating electricity.

This electricity is either directly sent to an electric motor or stored in batteries for later use. The electric motor then drives the wheels, making the car move. Unlike conventional cars, solar cars do not rely on an internal combustion engine, which means they produce zero tailpipe emissions.

The Evolution and Challenges

Early prototypes of solar cars were not necessarily practical for daily use. They were often slender, lightweight, and designed for maximum solar absorption rather than passenger comfort. However, modern solar cars are becoming increasingly consumer-friendly, with spacious interiors and amenities comparable to conventional vehicles.

But challenges persist:

- **Efficiency and Reliability:** Sunlight is variable. On cloudy days or during nighttime, solar cars rely on stored battery power. Ensuring consistent performance under all weather conditions remains a hurdle.

- **Infrastructure:** Charging infrastructure for electric vehicles is still developing in many parts of the world. For solar cars, this challenge is compounded by the need for specialized solar charging stations.

- **Cost:** Advanced photovoltaic materials and high-capacity batteries contribute to the high initial cost of solar cars.

Solar Bikes: Riding into the Future

While solar cars are garnering much attention, solar bikes represent another facet of the renewable mobility spectrum. These bikes use smaller solar panels, often attached to their wheels or frames, to provide an electrical boost to the rider.

Mechanics of Solar Bikes

Much like their four-wheeled counterparts, solar bikes rely on photovoltaic cells. The electricity generated either assists the rider directly or charges a battery. In the latter case, riders can use this stored energy to get an electric boost, especially useful for uphill rides or when one needs a break from pedaling.

The Appeal and Limitations

Solar bikes blend the benefits of electric bikes and the sustainable promise of solar energy. They offer a clean mode of transport, ideal for short distances or urban commuting. Moreover, the ability to generate one's own electricity on the go provides a sense of independence.

However, the limitations of solar bikes mirror those of solar cars on a smaller scale. The efficiency of the panels, reliance on weather conditions, and the initial cost can deter potential users.

Solar Boats and Drones

With over 70% of our planet covered in water, boats play an indispensable role in transport, recreation, and even residence for some. As environmental concerns intensify, the boating industry is reevaluating its traditional reliance on diesel and gasoline. Solar boats have emerged as a sustainable alternative, harnessing sunlight to power journeys across water.

The Concept of Solar Boating

Solar boats integrate photovoltaic panels, usually laid out on the deck or roof, to convert sunlight into electricity. This energy either directly powers electric motors or gets stored in onboard batteries for subsequent use. Since these boats don't rely on fuel combustion, they eliminate carbon emissions and potential oil spills, thus protecting marine ecosystems.

Why Solar Boats Make Sense

Water bodies, especially vast open seas and lakes, receive uninterrupted sunlight for long durations daily. This provides an ideal setting for harnessing solar energy. Additionally, the silence of electric motors ensures a tranquil sailing experience, preserving the serenity of aquatic environments.

However, there are challenges:

- **Power and Speed:** While advancements in technology are continually improving, the current generation of solar boats typically offers slower speeds compared to their diesel counterparts.

- **Battery Storage:** Prolonged sailing, especially during nights and cloudy days, necessitates efficient and high-capacity batteries, which can influence the vessel's weight and cost.

Solar Drones: A Skyward Ambition

Unmanned aerial vehicles (UAVs), commonly known as drones, have become ubiquitous in recent years, serving diverse functions from aerial photography to delivery services. Powering these drones sustainably is becoming increasingly pertinent, and solar energy presents a promising solution.

How Solar Drones Operate

Solar drones embed thin and lightweight solar panels on their wings or bodies. Sunlight absorbed by these panels gets converted into electricity, which then powers the drone's propellers and systems. Some advanced models can also store excess energy in batteries, enabling them to fly during non-daylight hours.

Potential and Implications of Solar Drones

The most significant advantage solar drones offer is prolonged flight durations. Traditional battery-powered drones have limited flight times, often necessitating frequent landings for recharges. With solar capabilities, drones can stay airborne for extended periods, making them ideal for tasks like surveillance, environmental monitoring, and even internet service provision in remote areas.

Yet, as with most solar technologies, challenges exist:

- **Panel Weight and Efficiency:** Striking a balance between drone mobility and the weight of the solar panels is crucial. The efficiency of these panels under varying weather conditions is another concern.

- **Initial Costs:** Incorporating solar technology can escalate the initial costs of drones. However, savings in energy costs over time can offset this.

Solar Power: Beyond Roads and Skies

Solar boats and drones are testaments to the versatility of solar energy. They extend the promise of renewable energy from our roads to our waters and skies.

As we sail and soar into a future marred by environmental uncertainties, these innovations remind us of the boundless potential of the sun. They underscore the need and the possibility to reimagine traditional systems, leveraging nature's bounty for sustainable progress. Each solar-powered voyage, whether over oceans or in the sky, signifies a step towards a brighter, cleaner world. And as technology continues to evolve, it's clear that the horizons of solar mobility are broad and full of promise.

CHAPTER 3: RVs, Vans, and Caravans

Installing Solar on Mobile Homes

The feeling of hitting the open road with the promise of new experiences around every turn is exhilarating. Recreational vehicles (RVs), vans, and caravans have long been the embodiment of this spirit of adventure. However, as these homes-on-wheels crisscross scenic highways and camp in remote locations, there's an increasing need to ensure they do so sustainably. Enter solar power, the champion of renewable energy, transforming the mobile home experience.

The Shift Towards Solar Mobile Homes

The very essence of a mobile home is freedom – freedom from being tied to a particular place, freedom to explore, and ideally, freedom from utility grids. Integrating solar panels onto RVs and caravans ensures a self-sufficient power source, reducing dependency on noisy generators or external power hook-ups.

Practicality and the Promise of Uninterrupted Power

A sunlit day can charge a mobile home's batteries, allowing occupants to use electrical appliances, lights, and systems into the night. This stored energy is particularly useful in remote camping locations where traditional power sources are unavailable or in high demand.

Harnessing the Sun: The Installation Process

When considering solar for your mobile home, there are several considerations:

- **Assessment of Energy Needs:** Before embarking on the solar journey, one needs to analyze the energy consumption patterns. How many appliances do you run and for how long? What's the combined energy draw?

- **Choosing the Right Solar Panels:** Flexible solar panels are increasingly popular for mobile homes due to their lightweight nature and adaptability to curved surfaces. However, rigid panels, while bulkier, often offer better durability and efficiency.

- **Battery Storage:** The captured solar energy needs storage for later use, especially during night-time or overcast days. Therefore, investing in a robust battery system is paramount. Modern lithium-ion batteries are favored for their longer lifespans and efficient energy storage capacities.

- **Inverter Selection:** An inverter converts the direct current (DC) from solar panels and batteries into alternating current (AC) for use in standard appliances. The size and type of inverter chosen should align with the energy requirements of the mobile home.

- **Monitoring and Maintenance:** Once the system is up and running, routine checks are essential. Modern solar setups often come with monitoring systems that provide real-time data on energy production and consumption, enabling timely interventions if issues arise.

Economic and Environmental Impacts

While the upfront costs of solar installations on mobile homes can be substantial, the long-term savings are noteworthy. With reduced fuel costs for generators and minimized fees at powered campgrounds, the return on investment becomes palpable over time.

Beyond economics, the environmental implications are profound. A solar-powered mobile home substantially reduces the carbon footprint of road-tripping, aligning the joy of travel with the conscientiousness of sustainable living.

Energy Management on the Road

There's an inherent romance in the concept of living on the open road, particularly in a vehicle that's both home and transport. But with this dual role comes dual responsibility: to ensure the journey is smooth and to guarantee a comfortable living environment. As more travelers embrace solar energy for their mobile homes, the challenge shifts from merely generating power to managing it efficiently.

Understanding Consumption

A nuanced understanding of how energy is consumed within a mobile home is foundational to effective energy management. Whether it's the refrigerator keeping perishables fresh, the heater combating cold nights, or the devices we depend on for navigation and entertainment, every appliance and gadget has its draw on the battery reserves.

Strategies for Efficient Energy Management

1. Prioritization: While it's tempting to use all appliances simultaneously, it's essential to prioritize. Understanding which devices are critical and which are luxury can help regulate energy use. For instance, while it might be appealing to use the microwave, heater, and charge devices at the same time, it may not always be feasible without draining the battery.

2. Efficient Appliance Choices: Investing in energy-efficient appliances can make a significant difference. LED lighting, for example, consumes a fraction of the energy of traditional bulbs. Similarly, modern refrigerators designed for mobile homes are optimized for low power consumption.

3. Regular System Checks: Like any other system, solar setups in mobile homes require regular maintenance to ensure they're working at optimal capacity. This involves cleaning solar panels to remove dust or debris, ensuring connections are secure, and periodically checking the health of batteries.

4. Optimal Parking: While on the move, it's beneficial to park in sunlit areas during the day to maximize solar energy collection. However, if the destination is hot, it's a balancing act between finding shade to keep the vehicle cool and ensuring the solar panels receive adequate sunlight.

5. Smart Monitoring Systems: With advancements in technology, several solar systems now come with sophisticated monitoring setups. These systems provide real-time data about energy generation, storage levels, and consumption patterns, allowing travelers to adjust their energy usage accordingly.

6. External Power Sources: Despite best efforts, there may be situations where solar energy might not suffice, especially during prolonged periods of overcast weather. In such instances, having a backup plan, such as a secondary battery or an energy-efficient generator, can be a lifesaver.

Energy management is more than a one-time setup; it's an ongoing process. As technology evolves, newer, more efficient devices come into the market, and consumption patterns change. Those living the mobile life must adapt continuously, learning from each trip and iterating their energy strategies.

CHAPTER 4: Solar in Disaster Relief

Emergency Solar Solutions

When calamity strikes, be it a natural disaster like hurricanes, floods, earthquakes, or man-made conflicts, one of the most immediate challenges is restoring essential services. Among these, power ranks high. Not just for the sake of comfort, but electricity is vital for emergency services, communication, and medical care. In such moments of desperation and chaos, solar energy emerges as an incredible ally, providing a reliable, quick, and sustainable power source.

The Immediate Aftermath: Powering Up When The Grid Goes Down

Post-disaster, traditional electrical infrastructures often suffer significant damage. It could take days, weeks, or even longer to restore power, especially in remote areas. In such situations, portable solar solutions can be the first responders of the energy world.

Portable Solar Generators: Compact and robust, portable solar generators are increasingly becoming a staple in emergency response kits worldwide. Unlike fuel-driven generators, these devices don't rely on a potentially disrupted fuel supply chain. They're silent, produce no emissions, and can be set up anywhere there's sunlight. These generators can power essential appliances like communication devices, lights, and medical equipment.

Solar Lanterns and Flashlights: In the absence of grid power, nights can become a daunting challenge, especially in disaster-stricken zones. Solar-powered lanterns and flashlights are simple solutions that can provide much-needed illumination. These devices usually come with built-in photovoltaic cells and can provide light for hours after a full charge.

Mobile Charging Stations: Communication is critical during emergencies. Keeping phones and radios charged ensures that those affected can contact loved ones and receive vital information. Portable solar chargers can help in these scenarios, offering a lifeline to the outside world.

The Sustenance Phase: Meeting Basic Needs Through Solar

Once the initial emergency phase has passed, the focus shifts towards providing sustained support to the affected communities. This involves ensuring a steady supply of clean water, food, and shelter, all of which can be enhanced using solar energy.

Solar Water Heaters and Purifiers: Access to hot water can be essential for hygiene and medical purposes. Solar water heaters, which use the sun's heat to warm water, are a great solution. Moreover, solar-powered water purifiers can utilize ultraviolet (UV) light to disinfect water, ensuring safe drinking sources for affected populations.

Solar Ovens and Cookers: Traditional cooking methods might be infeasible post-disaster, especially if there's a fuel shortage. Solar ovens, which concentrate sunlight to cook food, can be invaluable. They offer a smoke-free and fuel-free cooking alternative, reducing health risks and dependency on dwindling resources.

Temporary Solar-powered Shelters: As displaced people look for temporary shelters, tents and makeshift homes equipped with solar panels can offer some semblance of normalcy. These shelters can provide basic lighting and charging points, helping individuals stay connected and safe.

The Long-Term Perspective: Building Resilience with Solar

As communities recover and rebuild, it's crucial to learn from the experience and develop more resilient infrastructures. Solar microgrids, for example, can provide localized power solutions, ensuring that if one area is affected, others remain powered up. Additionally, promoting household solar solutions can help decentralize power generation, making communities less vulnerable to grid failures.

Portable Solar Water Purifiers

Water is the essence of life. Yet, in the aftermath of disasters, access to clean drinking water becomes a pressing concern. Contaminated water sources can lead to a host of waterborne diseases, further exacerbating the situation and turning it into a secondary crisis. In such dire scenarios, portable solar water purifiers emerge as a boon, addressing the immediate need for safe drinking water.

Harnessing Sunlight to Ensure Safe Drinking Water

The sun, with its abundant energy, can be harnessed not only to produce electricity but also to purify water. Solar water purifiers employ various technologies to ensure the water is free from pathogens and contaminants.

Solar Distillation: Distillation is one of the oldest methods of purifying water. Solar distillers utilize sunlight to heat untreated water. As the water evaporates, it leaves contaminants behind. The water vapor then condenses on a cooler surface, forming pure water droplets that are collected in a separate container. The resulting water is devoid of impurities, making it safe for drinking.

Solar Disinfection (SODIS): This simple method involves filling transparent plastic bottles with water and placing them in direct sunlight for several hours. The ultraviolet rays from the sun destroy harmful pathogens present in the water. While this method might not remove chemical contaminants, it's effective against bacteria, viruses, and parasites.

Photocatalytic Disinfection: This advanced technique uses the sun's UV rays in conjunction with a catalyst, usually titanium dioxide (TiO_2), to purify water. When the catalyst is exposed to UV light, it produces reactive oxygen species that can destroy organic contaminants and pathogens.

Solar-powered Electrochlorination: In this method, solar panels provide the power needed to produce chlorine by running a current through salty water.

This chlorine can then be used to disinfect larger volumes of water, making it safe for consumption.

The Far-reaching Impacts of Solar Water Purifiers

Immediate Relief: In the direct aftermath of a disaster, solar water purifiers can be deployed rapidly, providing affected communities with a reliable source of clean drinking water. This immediate access can prevent the onset of waterborne diseases, a common menace in such situations.

Empowerment of Displaced Populations: Refugees and internally displaced persons often reside in camps with limited infrastructure. Here, portable solar purifiers can offer a decentralized solution, allowing individuals and families to cater to their drinking water needs without depending on external aid continually.

Promoting Health in Remote Locations: Beyond emergencies, solar water purifiers hold promise for remote areas with limited access to clean water infrastructure. By providing communities with the tools to purify water using the sun, it becomes possible to reduce the incidence of waterborne diseases and improve overall community health.

Educational Opportunities: The introduction of solar water purifying technologies, especially in schools and community centers, can serve as a platform for broader educational initiatives. It presents an opportunity to teach children and adults alike about the importance of clean water, the science behind these purifiers, and the broader potential of solar energy.

Toward a Brighter, Healthier Future

As the world grapples with the dual challenges of climate change and increasing humanitarian crises, solutions that address multiple issues simultaneously are invaluable. Portable solar water purifiers represent such a solution.

By providing clean water, they not only avert health crises but also underscore the versatility of solar energy.

BOOK 9

Solar Maintenance And Troubleshooting

CHAPTER 1: Routine Solar Maintenance

Cleaning and Care

One might think that once a solar panel installation is complete, the system will function optimally without interference. While solar panels are indeed low-maintenance, they are not entirely maintenance-free. Like any other piece of technology, they require care to function at peak efficiency. A significant aspect of this care revolves around keeping them clean.

Dirt and Debris Impact on Efficiency

Dust, pollen, bird droppings, leaves, and other debris can accumulate on the surface of solar panels. These layers of grime can block sunlight and reduce the amount of solar energy that panels can absorb. Even a thin layer of dust can decrease the system's efficiency significantly.

In regions where sandstorms are common, sand particles can settle on the panels, exacerbating the decrease in efficiency. Likewise, in areas with heavy snowfall, accumulated snow can obscure the panels, rendering them temporarily non-functional.

The Process of Cleaning Solar Panels

Solar panels are delicate, and cleaning them requires care to avoid causing damage or scratches. Here's a general approach to cleaning solar panels:

1. *Safety First:* Always ensure that safety precautions are taken. This includes switching off the solar system before cleaning and using harnesses or safety ropes if the panels are located in elevated positions.

2. *Soft Cleaning:* Using a soft brush or a squeegee with a plastic blade, gently brush off loose dirt and debris. A long handle can be attached to the cleaning tool to reach panels that are mounted higher up.

3. Mild Detergents: If the panels are excessively dirty, use lukewarm water with a mild detergent. Avoid using abrasive materials or strong chemicals that can harm the panel's surface.

4. Rinsing: After cleaning with a detergent solution, it's crucial to rinse the panels thoroughly with clean water to remove any soap residues.

5. Avoiding Hard Water: In areas with hard water, it's recommended to use distilled or deionized water for the final rinse to prevent mineral deposits from forming on the panels.

Frequency of Cleaning

The required frequency of cleaning varies depending on the local environment. In dusty areas or places with heavy industrial activity, more frequent cleaning might be necessary. Conversely, in regions with regular rainfall, nature might assist in the cleaning process, reducing the need for manual intervention. However, even in such areas, an occasional check is advisable to ensure that stubborn grime or bird droppings aren't impacting the panel's performance.

Professional Cleaning Services

For those who aren't comfortable cleaning their panels or have installations that are hard to reach, professional cleaning services are available. These experts possess the necessary equipment and knowledge to clean the panels without causing damage. Moreover, they can often spot and rectify minor issues during their visits, ensuring the system remains in optimal shape.

Protecting Panels from Potential Damage

While cleaning is an integral part of maintenance, protecting the panels from potential damage is equally vital. Installations in areas prone to hail might benefit from protective measures like mesh screens.

Periodic System Checks

While the solar panel system inherently promises longevity and durability, no technology is entirely devoid of wear and tear. Therefore, regular inspections and system checks are paramount to ensure optimal performance over the years. Periodic checks can uncover potential issues before they escalate, saving both time and money in the long run.

Elements of a Solar System Check

Visual Inspections: At its simplest, a system check can be a visual inspection. Regularly observing the panels can reveal obvious issues such as physical damage, shading from new obstructions, or visible wear and tear. For instance, cracks, discoloration, or warping of the panels should prompt a deeper examination.

Checking the Mounting Equipment: Over time, the mounting equipment that holds your panels in place can loosen due to various factors like extreme weather or natural degradation of materials. It's essential to ensure that all panels are firmly attached and that there are no signs of corrosion in the mounting hardware.

Examine Electrical Components: While the panels themselves are crucial, the associated electrical components such as wiring, inverters, and connectors play an equally vital role in the system's performance. Check for signs of fraying, wear, or rodent damage in the wiring. Ensure that connectors are tight and show no signs of corrosion.

Inverter Monitoring: The inverter is the heart of your solar system, converting the DC power generated by the panels into usable AC power. Most modern inverters have monitoring systems that show how much power you're generating. A sudden drop in this output or frequent, unexplained fluctuations could indicate an issue.

Performance Monitoring: Several solar panel setups come with performance monitoring systems. These systems allow homeowners to track their system's daily, monthly, and yearly energy production. By keeping a close eye on this data, you can detect potential inefficiencies. For example, if you notice a consistent drop in energy production during what should be peak sunlight hours, there might be an obstruction or a panel malfunction.

Scheduled Professional Inspections: Even if you're diligent about checking your system, there's no substitute for a professional's trained eye. It's advisable to have a professional inspection done annually or biannually. Solar technicians can conduct in-depth checks, test component functionalities, and identify issues that might not be evident to the untrained eye.

Maintenance after System Checks

If, during these checks, you identify any issues or potential problems, it's crucial to address them promptly. Minor issues like a loose connector or a small shading obstruction can be fixed with little effort. However, larger problems such as malfunctioning inverters or degraded panels might require expert intervention.

The Role of Warranties

Many solar panel manufacturers provide long-term warranties, sometimes spanning 25 years or more. These warranties can be invaluable if you detect a malfunctioning panel or component. Keep all warranty information handy, and in case of issues, check if the affected component is still under warranty before paying for repairs or replacements.

CHAPTER 2: Diagnosing Common Issues

Panel Degradation

Solar panels, like any other technological equipment exposed to the elements, undergo degradation over time. This process, while natural and expected, can affect the efficiency and output of the panels. However, understanding panel degradation, its causes, and how to mitigate its effects can prolong the life of a solar system and ensure it operates at optimal levels for as long as possible.

What Exactly is Panel Degradation?

Panel degradation refers to the gradual decline in the power output of a solar panel over its operational life. Typically measured as a percentage, the degradation rate indicates how much less energy a panel will produce in a given year compared to the year before. For instance, a panel with a 0.5% annual degradation rate will produce 99.5% of its original output in its second year, 99% in its third, and so on.

Factors Contributing to Degradation

Several factors can influence the rate and extent of solar panel degradation:

1. Ultraviolet (UV) Exposure: Just as UV rays can cause harm to our skin, they can also affect the materials that solar panels are made from. Over time, prolonged UV exposure can break down the protective layers of a solar panel, potentially leading to issues such as delamination or yellowing.

2. Temperature Fluctuations: Extreme temperature changes, especially rapid fluctuations, can cause solar panel materials to expand and contract. Over time, this constant change can lead to micro-cracks in the cells or damage to the solder bonds.

3. Moisture: If water infiltrates the panel due to a compromised seal or a defect, it can lead to corrosion of the metal contacts and connectors. Moisture intrusion can also cause the solar cells themselves to degrade at an accelerated rate.

4. Physical Stress: Events like hailstorms, wind-blown debris, or even the weight of snow can cause physical damage to panels. Such physical stressors can create micro-cracks or even break cells, leading to reduced efficiency.

Detecting Panel Degradation

Visual Inspection: One of the first signs of panel degradation is often visible damage. Discoloration, yellowing, or fogging within the panel can be signs of UV or moisture damage. Physical deformities like warping, delamination, or visible cracks are also indicative of degradation.

Performance Monitoring: By consistently monitoring the energy output of a solar system, you can detect signs of degradation. If a system's output begins to decline outside of expected efficiency losses, it's an indication that one or more panels might be degrading at a faster rate.

Using Electroluminescence Imaging: Professionals often use electroluminescence (EL) imaging to detect defects and degradation in solar panels. EL imaging involves applying a voltage to the panel and capturing the emitted light patterns. Variations in these patterns can reveal defects like micro-cracks or broken cells.

Addressing Panel Degradation

The approach to addressing panel degradation largely depends on its severity and cause. Minor degradation or wear might only require a thorough cleaning or resealing of the panel's edges. More severe degradation, especially if it impacts the panel's functionality, might necessitate a replacement.

When replacing panels due to degradation, it's vital to consider compatibility. Newer panels might have different efficiency levels or electrical characteristics than older ones. Therefore, when integrating new panels into an existing system, it's crucial to ensure they're compatible to prevent potential imbalances or inefficiencies.

Inverter and Battery Failures

Inverters are subjected to constant stress due to the cyclical nature of solar production. They experience surges in power during peak sun hours and periods of inactivity during low-light conditions.

Causes of Inverter Failures:

1. **Electronic Component Degradation:** Inverters contain numerous capacitors, switches, and other electronic components that can degrade over time. Heat, which is a byproduct of the inversion process, can expedite this degradation.

2. **Software or Firmware Issues:** Inverter operations rely on firmware to regulate and manage the conversion process. Outdated or glitch-prone software can lead to operational hiccups.

3. **External Factors:** Factors like moisture ingress, pests, or even dust and debris can affect an inverter's functionality. Properly sealed enclosures and routine checks can mitigate these risks.

Solutions for Inverter Failures:

- **Regular Maintenance:** Ensuring that the inverter is clean and its cooling system is working efficiently can prolong its lifespan. This includes cleaning the inverter's filters and ensuring adequate ventilation.

- **Firmware Updates:** Regularly updating the inverter's software can address known bugs and improve its performance.

- **Professional Diagnosis:** If an inverter consistently underperforms or shows error messages, it might be time to call in a professional to diagnose and potentially replace it.

Battery Failures: Causes and Solutions

Solar batteries, especially when used in daily-cycling systems, undergo significant wear and tear. Over time, this can reduce their capacity and efficiency.

Causes of Battery Failures:

1. **Deep Discharges:** Frequently discharging a battery to its lowest capacity can reduce its overall lifespan. This is especially true for certain types of batteries, like lead-acid.

2. **Temperature Extremes:** Batteries are sensitive to temperatures. Extremely cold or hot conditions can impair their performance and longevity.

3. **Poor Maintenance:** Some batteries, particularly flooded lead-acid ones, require regular maintenance, including topping off with distilled water.

Solutions for Battery Failures:

- **Temperature Management:** Keeping batteries in a temperature-regulated environment can drastically improve their efficiency and lifespan. Some modern battery systems come with built-in thermal management systems.

- **Depth of Discharge Management:** Utilizing battery management systems that prevent deep discharges can prolong battery life. Understanding the optimal depth of discharge for a specific battery type is crucial.

- **Regular Checkups:** Periodic checks can reveal issues like swelling, leakage, or other signs of failure. Addressing these early can prevent more severe damage.

While solar panels often capture the limelight when discussing solar energy, inverters and batteries are equally crucial to the system's functionality. Regular maintenance, along with a proactive approach to identifying and resolving issues, ensures that a solar energy system remains robust and efficient throughout its operational life.

CHAPTER 3: Upgrading and Expanding

Adding More Panels

Adding more panels to an already established setup is a common method of achieving this expansion, but it's not as simple as just bolting on more modules. There are various considerations, challenges, and benefits associated with this endeavor.

The Rationale Behind Expanding Solar Arrays

Solar panel expansion typically stems from two primary motivators:

1. **Increased Energy Consumption:** Changes in lifestyle, growing families, or expanded business operations might elevate energy demands. For instance, introducing electric vehicles, new machinery, or additional electronic appliances can substantially boost energy requirements.

2. **Desire for Greater Energy Independence:** Many individuals initially adopt solar energy due to environmental concerns or financial benefits. Over time, they may wish to become entirely energy independent, reducing or eliminating their reliance on grid power.

Factors to Consider When Adding Panels

1. **Assessment of Energy Needs:** Before considering an expansion, it's essential to determine the additional energy requirements. Monitoring tools and energy audits can provide insights into current consumption patterns and project future needs.

2. **Space Availability:** Additional panels necessitate extra space. Rooftop installations might be constrained by the existing structure, requiring homeowners to explore ground-mounted options or alternative placements.

3. **Compatibility:** Not all solar panels are created equal. Mixing old panels with new ones can lead to mismatches in voltage, efficiency, or power output. This discrepancy can impact the overall performance of the solar array. In such situations, using power optimizers or micro-inverters can help synchronize varying panel outputs.

4. **Inverter Capacity:** The existing inverter's capacity is paramount. If the inverter can't handle the increased output from additional panels, it might require an upgrade or the addition of a secondary inverter.

5. **Regulatory and Permitting Issues:** Local regulations, utility interconnection standards, and permitting processes might have changed since the initial installation. It's vital to be well-acquainted with current requirements before proceeding.

Steps to Adding More Panels

- **Professional Consultation:** Begin with consulting solar installation experts. They can assess the existing system, evaluate the structural integrity of the installation site, and suggest optimal expansion strategies.

- **Budgeting and Financing:** With an expansion plan in hand, it's easier to budget for the project. Many financing options, such as loans, leases, or power purchase agreements, can help spread out the costs.

- **Acquisition and Installation:** Once the financial aspects are settled, procuring the panels, inverters, and other necessary components follows. Professional installers can then integrate the new panels with the existing setup.

- **System Testing:** After installation, it's crucial to test the expanded system thoroughly. This step ensures that all components work harmoniously and the system delivers the expected power output.

- **Monitoring and Maintenance:** An expanded solar system might necessitate updated monitoring solutions. Newer panels may come with advanced monitoring features, providing more granular data on system performance.

Benefits of System Expansion

1. **Higher Energy Output:** The most apparent benefit is the increased energy production, which can cater to growing power needs or even feed excess power back to the grid, depending on net metering policies.

2. **Enhanced Financial Savings:** More solar production means reduced dependence on grid electricity, leading to more significant financial savings. Additionally, some regions offer incentives or rebates for solar expansions, further enhancing ROI.

3. **Reduced Carbon Footprint:** By covering a more substantial portion of energy needs through solar, homeowners and businesses can further diminish their carbon footprints, contributing positively to environmental preservation.

Switching to Advanced Components

While solar panels can have operational lives extending up to 25 years or more, some system components might require earlier replacements or benefit from technological upgrades. These components may include inverters, batteries, mounting systems, and monitoring systems.

Several factors can influence the decision to upgrade:

1. **Technological Advancements:** Newer components often offer better efficiency, durability, and functionality. For instance, recent inverter models may have higher conversion efficiencies or provide enhanced features for monitoring and control.

2. **Wear and Tear:** Even with diligent maintenance, parts like inverters usually have a shorter lifespan (typically 10-15 years) compared to panels. Batteries, depending on the type and usage, might also degrade faster.

3. **Expanding System Capacity:** As discussed, adding more panels might necessitate upgrades in inverters or battery storage to handle the increased power output.

4. **Desire for Enhanced Features:** Modern components can offer features like better system monitoring, integration with smart home systems, or improved safety functionalities.

Key Components to Consider for Upgrades

- **Inverters:** These devices, which convert the direct current (DC) produced by solar panels into alternating current (AC) usable by home appliances, often see substantial technological advancements. Newer models might offer modular designs, allowing them to scale effortlessly with system expansion. Some modern inverters also incorporate AI-driven diagnostics, predicting maintenance needs or identifying efficiency bottlenecks.

- **Batteries:** Battery storage technology, especially with the rise of lithium-ion batteries, has seen rapid advancements in recent years. Newer batteries offer longer lifespans, faster charging, deeper discharge capabilities, and improved safety features.

- **Monitoring Systems:** The ability to track, analyze, and optimize solar system performance has become increasingly sophisticated. Upgrading to a modern monitoring solution can provide real-time insights, predictive maintenance alerts, and integration with other home or business management systems.

- **Mounting Systems:** Advanced mounting solutions may offer better durability, flexibility in panel positioning, or even tracking functionalities to adjust panel angles based on the sun's position.

The Upgrade Process

Upgrading solar components isn't as simple as swapping out old for new. It involves:

- **Evaluation and Consultation:** Before any upgrade, a thorough assessment of the current system is vital. This evaluation will identify the need for upgrades and ensure compatibility.

- **Cost Analysis:** Upgrades are investments. A clear understanding of costs, potential energy yield improvements, and ROI timelines will aid in making informed decisions.

- **Installation and Integration:** Once components are chosen, professional installation ensures they're integrated seamlessly into the existing system. Proper integration is crucial to harness the full benefits of the new technologies.

- **Testing and Calibration:** Post-upgrade, it's imperative to test the system comprehensively, calibrating any settings to ensure optimal performance.

Switching to advanced solar components is a strategic decision. While the initial costs can be substantial, the long-term benefits in efficiency, functionality, and durability often justify the investment.

CHAPTER 4: Ensuring Long-term Performance

Monitoring System Efficiency

Solar system monitoring serves as the diagnostic tool for solar installations. Just as a regular health check-up detects potential medical issues, a solar monitoring system provides real-time insights into the performance of a solar array. It pinpoints areas that are functioning optimally and highlights those that need attention.

Why Monitor?

The foremost reason to keep a close eye on solar system efficiency is to ensure that you're getting the most out of your investment. A slight dip in performance can indicate various issues, ranging from soiling on panels and shading to more technical problems like inverter malfunctions or wiring issues.

Beyond this, monitoring assists in:

- **Predictive Maintenance:** By tracking system performance, you can often predict when specific components might fail or require maintenance, allowing you to address problems before they escalate.

- **Optimizing Energy Consumption:** Many modern monitoring solutions not only show the energy production of your solar panels but also your household or business energy consumption. By analyzing these data, one can make informed decisions about energy use, optimizing it to coincide with peak solar production hours.

- **Financial Tracking:** Monitoring systems can often be set to track savings over time, calculating the financial benefits gained from solar energy generation.

Key Metrics to Monitor

- **Energy Production:** This is the measure of how much energy your solar panels are producing. It can usually be viewed on daily, monthly, or yearly timelines.

- **Energy Consumption:** As mentioned, some systems allow for tracking energy usage alongside production, giving a comprehensive view of energy dynamics in a particular setup.

- **System Health:** This includes metrics that flag system or component malfunctions, indicating the need for maintenance or repair.

- **Environmental Impact:** Many monitoring solutions calculate the carbon offset of your solar energy production, providing data on how much your solar panels are contributing to environmental sustainability.

Modern Monitoring Techniques

With the integration of artificial intelligence and machine learning, modern monitoring solutions have evolved from mere data display tools to comprehensive analytic systems:

- **Remote Monitoring:** Leveraging the internet, many modern systems allow homeowners or business owners to check their solar system's performance from anywhere in the world through smartphones or computers.

- **Predictive Analytics:** Advanced monitoring solutions can predict potential issues or forecast energy production based on historical data and weather predictions.

- **Integration with Other Systems:** With the rise of smart homes and businesses, solar monitoring systems often integrate with other building management systems, providing a holistic view of energy dynamics.

Challenges and Solutions

- **Data Overload:** With the wealth of information available through modern monitoring systems, it can be easy to become overwhelmed. It's crucial to focus on key metrics that directly impact system performance or financial returns.

- **Accuracy:** Ensuring the data accuracy of monitoring systems is vital. Regular calibrations and system checks can help in maintaining the fidelity of data readings.

- **Security:** As with all connected systems, there's a risk of cyber-attacks. Investing in monitoring solutions that prioritize security and regular software updates is essential.

Monitoring system efficiency stands as the linchpin in ensuring the longevity and optimal performance of a solar installation. By continually keeping tabs on the system's output and health, users can ensure they maximize their energy savings, contribute positively to the environment, and elongate the life of their solar setup.

Seasonal Adjustments and Checks

Every geographical location on Earth experiences variations in the sun's trajectory throughout the year. During summer, the days are longer, and the sun is usually higher in the sky, leading to increased solar energy production. Conversely, winter months bring shorter days, and the sun remains relatively lower, resulting in reduced solar output. Spring and autumn, being transitional seasons, offer moderate solar energy potential.

The Need for Seasonal Adjustments

Ensuring consistent solar energy production throughout the year requires adjustments based on these seasonal shifts. A solar system set for summer's high sun might not capture the winter sun efficiently, leading to a significant drop in energy production.

Adjusting Solar Panel Angles

One of the most effective methods to combat seasonal variations is by adjusting the tilt angle of solar panels. Solar panels generate the most energy when they are perpendicular to the sun's rays. By altering the tilt according to the season, one can ensure maximum sunlight exposure.

In regions with distinct seasonal changes, adjustable solar panel mounts can be a worthy investment. These allow the system's angle to be changed multiple times throughout the year. For instance, during summer, a lower tilt might be optimal, while in winter, a steeper angle might be more suitable.

Seasonal System Checks

Beyond adjusting panel angles, conducting regular seasonal system checks is essential. These checks can include:

- **Cleaning Panels:** As seasons change, so do environmental conditions. Autumn might bring fallen leaves, while winter could deposit snow or frost on panels. Regular cleaning ensures the panels remain unobstructed and efficient.

- **Inspecting Wiring and Connections:** Temperature fluctuations can affect the physical components of a solar system. Expansion and contraction due to varying temperatures might lead to loosened connections, warranting periodic checks.

- **Calibrating Monitoring Systems:** As the sun's trajectory changes, recalibrating monitoring systems ensures accurate performance tracking. It aids in identifying any deviations from expected energy production, indicating potential issues.

- **Checking Inverters and Batteries:** These components are sensitive to extreme temperatures. Ensuring they are well-protected and functioning efficiently is crucial, especially during the peak of summer or the depth of winter.

For solar systems with energy storage solutions, understanding seasonal energy consumption patterns is as important as knowing production trends. Energy demands often vary seasonally, with higher consumption in winters due to heating or in summers due to cooling. Properly calibrated energy storage ensures that there's always a backup, especially during periods of reduced solar production.

BOOK 10

Financing And Policies For Solar Adoption

CHAPTER 1: Understanding Solar Financing

New Solar Cell Materials and Their Potentials

Historically, crystalline silicon has been the dominant material for solar panels. Its abundant availability, mature manufacturing processes, and reasonable efficiency made it a go-to choice. However, silicon panels have some limitations, particularly concerning their weight, rigidity, and efficiency thresholds.

The relentless quest for higher efficiencies and flexible applications has ignited interest in alternative materials that could revolutionize solar energy.

Perovskite Solar Cells: A Beacon of Promise

One of the most talked-about materials in recent solar research is perovskite. These materials, characterized by their unique crystal structure, have demonstrated rapid efficiency improvements over a relatively short span.

The appeal of perovskite lies not just in its efficiency but also in its versatility. It can be engineered to harness different parts of the solar spectrum, making tandem solar cells, where perovskite is layered on top of silicon, a hot topic of research. This combination aims to utilize the best attributes of both materials, pushing efficiency limits beyond what's currently achievable with silicon alone.

Moreover, the possibility of crafting lightweight, semi-transparent, and flexible perovskite solar panels could redefine how we integrate solar technology into urban landscapes, transportation, and consumer electronics.

Quantum Dots: Harnessing Nanotechnology

At the intersection of quantum physics and solar research lie quantum dots – semiconductor particles mere nanometers in size. Their minuscule size gives rise to quantum mechanical properties, allowing for the tuning of the material's electronic characteristics.

Quantum dots have the potential to absorb different parts of the solar spectrum, much like perovskite, but with the added advantage of being tunable due to their quantum nature. This means engineers can adjust their properties to maximize efficiency based on specific applications or environments.

Organic Photovoltaic Cells (OPVs): Flexible and Lightweight

Organic solar cells, made from carbon-rich compounds, offer another avenue for innovation. Their primary appeal is their potential for flexibility, lightweight, and semi-transparency. Think of rollable solar panels or solar coatings on windows.

While their efficiencies are currently lower than their inorganic counterparts, ongoing research aims to bridge this gap, ensuring OPVs find their niche in specific applications where their unique properties are invaluable.

Bifacial Solar Cells: Capturing Light from Both Sides

Traditional solar panels are monofacial, meaning they capture sunlight from one side only. Bifacial solar cells, on the other hand, harness sunlight from both their front and rear sides. By capturing light reflected off surfaces or even direct sunlight during specific times of the day, they boost the overall energy yield. While not precisely a new "material," the design innovation offers another avenue to augment solar energy production.

Future Implications and Challenges

As promising as these materials sound, challenges remain. Issues like long-term stability, scalability of production, and integration with existing infrastructure are but a few of the hurdles researchers face.

However, the potential benefits are undeniable. Higher efficiencies mean more energy from the same surface area, reducing land use and maximizing energy yield in space-constrained areas. Flexibility and lightweight properties can revolutionize mobile applications and building-integrated photovoltaics.

Breakthroughs in Solar Cell Designs

The concept behind tandem solar cells is simple yet profound. By layering two (or more) solar cell materials with different band gaps, it's possible to capture a broader spectrum of sunlight. The idea is to have the top cell absorb the higher energy photons (like ultraviolet light) while allowing the lower energy photons (like infrared) to pass through and be absorbed by the bottom cell.

Perovskite-silicon tandem cells, as touched upon earlier, exemplify this approach. But there are other combinations being explored too, like gallium arsenide with silicon. The potential efficiency gains from tandem cells can push solar power into new frontiers, making it even more competitive with conventional energy sources.

Light-trapping Designs: Ensuring Every Photon Counts

A photon missed is energy wasted. Recognizing this, researchers have been looking into designs that can trap light within the solar cell, increasing the chances of absorption. Nanostructures, textured surfaces, and reflective coatings are some strategies employed to keep light bouncing within the cell until it's absorbed.

Such designs aim to squeeze out every bit of energy from incoming sunlight, especially beneficial in areas with less than optimal sunlight.

Heterojunction Cells: Bridging Different Worlds

Heterojunction technology (HIT) involves combining materials with different atomic arrangements, typically amorphous silicon with crystalline silicon. This marriage of materials results in unique electrical properties at their junction, reducing electron recombination (where electrons don't contribute to current but instead recombine with holes).

Fewer recombinations mean more electrons contribute to the electric current, thereby boosting efficiency. Panasonic, a leading electronics company, has been at the forefront of HIT technology, pushing its efficiencies to impressive levels.

Flexible and Thin-film Solar Cells: Redefining Form Factors

Not all applications require rigid, bulky panels. For integration into wearables, portable electronics, or even certain architectural elements, flexibility is key. Thin-film solar cells, made by depositing one or more thin layers of photovoltaic material on a substrate, offer this advantage.

Amorphous silicon, cadmium telluride, and copper indium gallium selenide are materials commonly used in thin-film technology. While they generally have lower efficiencies than traditional crystalline silicon cells, their potential for flexibility, lightweight, and even semi-transparency offers unique applications.

Concentrated Photovoltaics (CPV): Intensifying Sunlight

CPV systems, rather than trying to modify the solar cell's intrinsic properties, focus on an external strategy. They employ lenses or mirrors to concentrate sunlight onto a small, highly efficient solar cell. By intensifying the light, the solar cell can generate more power. However, CPV systems typically require tracking mechanisms to stay aligned with the sun and are best suited for areas with high direct sunlight.

CHAPTER 2: Maximizing Solar Potential

Expanding Solar Farms and Harnessing Desert Power

In our insatiable quest for energy, we have often looked at our planet's vast landscapes, searching for spaces where we can harness the immense power of the sun. The untapped potential of large, sun-drenched expanses, especially deserts, presents a tantalizing proposition for the world of solar energy.

Deserts: Nature's Solar Goldmine

Deserts, with their wide-open spaces, minimal cloud cover, and intense sunlight, are nature's goldmine for solar energy. These arid regions receive solar irradiation levels that far exceed global averages. To put it in perspective, the Sahara Desert alone, if fully covered with solar panels, could produce more electricity than the entire world consumes. But beyond just potential, transforming these barren landscapes into energy hubs could also bring socio-economic benefits to often marginalized desert communities.

The Challenge of Transmission and Infrastructure

Harnessing desert power isn't just about placing solar panels on sand. The infrastructure needed to collect, store, and most importantly, transmit this energy to where it's needed is monumental. Deserts, being remote and vast, pose logistical challenges. Setting up transmission lines from desolate desert areas to bustling urban centers, which might be thousands of kilometers away, involves huge costs and technical challenges. There's also the issue of energy loss during transmission over long distances, which would need advanced solutions like high-voltage direct current (HVDC) transmission systems.

The Mirage of Water Needs

Solar farms, contrary to what some might think, do require water, especially for cleaning panels to maintain efficiency. In a desert, water is, of course, a scarce resource. Hence, innovative water-saving technologies or alternative cleaning methods would be paramount. Some companies are already exploring robotic cleaners and advanced coatings to minimize dust accumulation, thereby reducing water needs.

Socio-economic Impacts: A Double-edged Sword

While solar farms can bring jobs and infrastructure development to desert regions, it's crucial to approach these projects with sensitivity to local ecosystems and communities. Massive solar installations could disrupt local wildlife habitats, and without careful planning, might not always equate to long-term, sustainable jobs for local residents.

Desertec: A Glimpse into the Future

One of the most ambitious projects in this realm was the Desertec initiative. Envisioned as a massive interconnected grid, it aimed to tap into the solar and wind potential of the Sahara and Middle Eastern deserts to power a significant portion of Europe. While the project has faced hurdles and evolved over the years, its very conception underscores the global interest in desert power.

Small-scale, Distributed Desert Solar Projects

While mega solar farms capture headlines, there's also merit in considering distributed, smaller-scale solar installations in desert regions. Such projects can be quicker to deploy, reduce transmission losses (since they cater to local energy needs), and can be more easily integrated with community development initiatives.

Beyond Just Solar: The Potential for Integrated Energy Hubs

Deserts, with their vast spaces, are not only suitable for photovoltaic panels but also other renewable technologies. Concentrated solar power (CSP) that uses mirrors or lenses to focus sunlight, producing heat that drives turbines, is another viable desert technology. Combining CSP with photovoltaics can lead to hybrid systems that produce power round the clock – PV for the day and CSP, with its energy storage capabilities, for the night.

Urban Rooftop Solar: Brightening the Concrete Jungle

Amidst the towering skyscrapers and bustling streets, urban landscapes might seem an unlikely frontier for solar energy. Yet, the expansive rooftops that span across cities globally present an underutilized asset, ready to be transformed into hubs of green energy. Urban rooftop solar, while not a new concept, is gaining renewed interest as cities grapple with rising energy demands, environmental concerns, and the push for sustainability.

Concrete Jungles with Solar Canopies

The urban environment, with its dense concentration of buildings, offers a plethora of flat surfaces ideal for solar installations. From residential homes to vast commercial complexes, every sunlit rooftop is a potential mini power plant. Cities, already consuming a large chunk of the world's energy, have an opportunity to produce a significant portion of their electricity right where it's needed, minimizing transmission losses.

Navigating the Urban Maze

Installing solar panels in urban settings isn't without challenges. Rooftops, especially in older buildings, might not always be structurally sound to support solar installations. Then there's the issue of shadows. In densely packed cities, surrounding buildings, infrastructure, and even trees can cast shadows, impacting the efficiency of solar panels. Dynamic solar assessments, leveraging technology and data analytics, are thus crucial to ascertain the feasibility and potential output of urban solar installations.

Economic and Policy Drivers

The falling costs of solar panels combined with various incentives offered by municipalities and governments make urban solar economically attractive. In several cities, homeowners can benefit from net metering, where excess electricity generated is fed back into the grid, effectively turning the meter backwards. Such policies not only make solar installations financially viable but also reduce the strain on the central grid, especially during peak hours.

Community and Shared Solar Initiatives

Not every urban resident has access to a suitable rooftop, especially in cities with a high prevalence of apartment living. Enter community solar – a model where multiple individuals or entities come together to invest in and benefit from a shared solar installation. Such projects can be situated on larger communal buildings, unused urban spaces, or even outside the city limits. Participants receive credits on their utility bills corresponding to their share of the solar energy produced.

Integrating with Urban Infrastructure

Beyond just rooftops, there's potential for integrating solar panels into various facets of urban infrastructure. Solar bus stops, benches with charging stations, and solar-powered street lights are just a few examples. As urban planners and architects embrace sustainable design principles, we can expect a seamless fusion of solar technology into the very fabric of our cities.

Resilience Against Grid Failures

Urban centers are particularly vulnerable to power outages, given the high dependency on electricity for everything from transportation to basic services. Distributed solar installations across the city, especially those paired with energy storage solutions, can offer a buffer against grid failures. In the event of natural disasters or other disruptions, having multiple decentralized power sources can be a game-changer for urban resilience.

CHAPTER 3: Solar Policies and Regulations

Redefining Cityscapes: Solar Streets and Public Spaces

The concept of integrating solar technology into city streets and public spaces represents a visionary step in urban planning. The idea goes beyond merely placing solar panels on rooftops or in remote solar farms. Instead, it's about weaving the power of the sun into the very fabric of urban infrastructure. In this pursuit, streets, parks, and plazas are no longer just static entities but dynamic elements that play an active role in energy production.

Harnessing Energy Beneath Our Feet

In most urban settings, roads and pavements cover a substantial portion of the land. So, the idea of turning these vast stretches into energy generators seems both ambitious and logical. Solar roads, for instance, involve embedding photovoltaic cells within the road's surface. These specially designed panels are robust, capable of withstanding the weight of vehicles, and provide a slip-resistant surface similar to conventional tarmac. But beyond these physical attributes, they are efficient energy converters, turning the sun's rays into electricity even on overcast days.

Solar Sidewalks and Plazas

While the idea of solar roads is making headway, an even more intriguing prospect is that of solar sidewalks and plazas. These pedestrian areas don't have to contend with the weight of heavy vehicles, allowing for a broader range of materials and designs. Transparent solar tiles, for instance, can be utilized, letting light through and creating aesthetically pleasing patterns. As people stroll, shop, or simply relax, the ground beneath them quietly generates power.

Incorporating Smart Technology

The integration of solar technology in streets and public spaces provides an avenue for cities to become smarter. Embedded sensors within solar roads or pavements can gather a plethora of data – from monitoring traffic patterns and pedestrian movement to detecting any structural issues. This data can be invaluable for urban planning, enhancing safety, and optimizing traffic flow. Furthermore, these solar pathways can also incorporate features like LED lights that illuminate the roads at night, improving visibility and safety.

Powering Public Amenities

Urban spaces are adorned with numerous amenities that require power. Think of streetlights, public Wi-Fi hubs, charging stations, digital signages, and even water fountains. Typically, these are powered by the main grid. However, with solar-integrated streets and plazas, there's an opportunity to make these amenities self-sufficient. For instance, a solar-powered bench in a park could offer USB charging ports. A solar plaza could provide power to a surrounding outdoor cafe, allowing it to operate without drawing power from the grid.

Economic and Social Implications

The financial outlay for transforming urban surfaces into solar entities is significant. Yet, the returns, both direct and indirect, are substantial. Direct returns come from the energy generated, which reduces the reliance on external power sources, leading to cost savings. Moreover, cities can generate revenue by feeding excess energy back into the grid.

On the social front, solar streets and spaces can be instrumental in educating the public about renewable energy. Interactive installations, which allow people to see the amount of energy generated in real-time, can be both informative and motivational.

Despite the promise, there are challenges to be addressed. The initial cost of installation is a primary concern. Solar roads, for instance, are currently more expensive than traditional roads. There's also the matter of durability and maintenance. Roads are subjected to significant wear and tear, and ensuring that the embedded solar panels remain functional over the years is paramount.

Technological advancements are gradually addressing these challenges. Research is ongoing to develop materials that are both cost-effective and durable. Moreover, pilot projects across various cities worldwide are providing valuable insights, shaping the future trajectory of this venture.

Beyond the Horizon: Solar Skyscrapers and Vertical Integration

Skyscrapers, by their very nature, provide vast vertical surfaces that remain exposed to the sun for extended periods, especially in densely populated cities where taller buildings are less likely to be overshadowed by their neighbors. This unique architectural design makes them prime candidates for solar integration.

Incorporating BIPV (Building Integrated Photovoltaics)

BIPV technology allows for the integration of photovoltaic materials directly into building structures, without the need for additional structures or alterations. This means that windows, facades, and even outer walls can be constructed using materials that simultaneously serve architectural and energy-generating purposes. Unlike traditional solar panel installations, BIPV installations are more aesthetically pleasing, as they blend seamlessly into the building's design. This factor alone has accelerated its popularity among architects and city planners. An additional benefit of BIPV is that it can be incorporated during the construction phase, reducing the complexities and costs of retrofitting existing structures.

Balancing Transparency with Efficiency

One of the most exciting developments in vertical solar integration is the advent of transparent solar panels. These panels, which can be integrated into windows, allow skyscrapers to maintain their transparent, glass-dominated facades while still harvesting solar energy. This advancement ensures that buildings do not have to sacrifice natural light, views, or aesthetic appeal for energy efficiency.

Thermal Benefits and Energy Efficiency

Beyond electricity generation, vertical solar integration offers thermal benefits. BIPV systems can act as an additional layer of insulation, reducing the building's heating and cooling needs. In regions with extreme temperatures, this dual functionality—energy generation and insulation—can significantly reduce a building's carbon footprint and operational costs.

Challenges in Vertical Integration

While the prospects are promising, challenges remain. First, the angle at which vertical surfaces receive sunlight is not always optimal for energy generation, meaning vertically integrated panels may not be as efficient as their rooftop counterparts. This issue becomes even more pronounced in cities at higher latitudes, where the sun remains low on the horizon for a significant portion of the year.

Furthermore, integrating solar technology into buildings, especially skyscrapers, requires coordination across various stakeholders, including architects, engineers, city planners, and regulatory bodies. The intricacies of building design, combined with the technicalities of solar integration, necessitate a multidisciplinary approach.

CHAPTER 4: The Future of Solar Policy

The Solar Renaissance: Harnessing Art and Culture

The interplay between renewable energy and cultural expressions paints a vivid picture of a society pivoting towards a greener future. As cities around the world grapple with the challenges posed by climate change, rising energy costs, and urbanization, a unique intersection has emerged, drawing together the threads of art, culture, and solar technology.

The connection between art and technology has a long history. From the intricate designs of ancient sundials to the modern, dynamic light installations powered by solar, art has often been a medium to reflect humanity's understanding of the sun. Today, the fusion of solar technology with art is not just symbolic; it is functional, aesthetic, and profoundly impactful.

The Aesthetics of Solar Installations

For too long, solar installations were seen as purely utilitarian, often with little regard for their visual appeal. This perspective has undergone a drastic shift. Artists, architects, and designers are increasingly viewing solar panels not just as energy generators, but as canvases that can be molded and integrated into artistic expressions.

One of the most prominent examples is solar sculptures—dynamic installations that not only generate electricity but also captivate onlookers. These sculptures, sometimes moving with the sun, can be found in public spaces, gardens, and museums, serving as a testament to the symbiotic relationship between form and function.

Integrating Solar in Historic Sites

Historical sites, with their rich tapestry of stories and architecture, might seem an unlikely place for modern solar integration. Yet, conservationists and solar enthusiasts have found common ground. To maintain these sites, energy is required, and what better way to ensure their preservation than by harnessing renewable energy?

The challenge lies in integrating solar technology without detracting from the historical value or aesthetics of the site. Customized solar tiles that mimic the appearance of traditional roofing materials or transparent solar installations that blend seamlessly with ancient glasswork are innovations that bridge the old with the new. Such initiatives underline a profound respect for the past while paving the way for a sustainable future.

Solar Festivals and Cultural Celebrations

Across the globe, festivals celebrating the sun have existed for millennia. Today, many of these celebrations have adopted a modern twist, incorporating solar technology. From solar-powered lanterns lighting up night skies during festivals to stages and sound systems powered by solar at music festivals, the blend of tradition and technology is evident.

Moreover, new festivals dedicated entirely to celebrating solar energy have sprung up. These events serve as platforms to educate, inspire, and showcase the latest innovations in the solar realm. They often blend music, art installations, workshops, and interactive sessions, offering attendees a holistic experience of the solar revolution.

Education through Artistic Solar Expressions

Art has an innate ability to communicate complex ideas in accessible ways. Solar-inspired art installations in public spaces or educational institutions serve as conversation starters, drawing people into discussions about renewable energy, its potential, and its importance.

Interactive solar installations, especially those in science museums or educational campuses, play a pivotal role in shaping young minds. When children and young adults can touch, see, and interact with solar-driven artworks, their understanding deepens, and the seeds for a greener future are sown.

As society becomes more conscious of its ecological footprint, the marriage between art, culture, and solar technology will only strengthen. Artists will continue to draw inspiration from the sun, shaping materials and technologies to reflect our evolving relationship with this life-giving star.

This Solar Renaissance goes beyond mere installations. It represents a collective awakening to the beauty and potential of solar energy. It serves as a reminder that technology, when harmoniously integrated with art and culture, has the power to transform landscapes, mindsets, and futures.

Solar-Powered Urban Edens: The Role of Parks and Green Spaces

Modern urban parks have begun to evolve from being just recreational spaces to multifunctional arenas that contribute to the city's sustainability. Solar power, with its ever-advancing technology, has found a significant role in this transformation.

Water features in parks, such as fountains and artificial waterfalls, can now be operated using solar pumps. These pumps store energy during daylight hours and can power the water features into the evening, creating picturesque settings without drawing on the city's power grid.

Similarly, pathways and jogging tracks are being illuminated with solar-powered lights. Not only do these reduce electricity costs, but they also ensure that the parks are safe and accessible even after sunset. These lights, often equipped with sensors, ensure energy is used judiciously—turning on only when someone is near and dimming down or switching off when not needed.

The Allure of Solar Canopies and Shelters

In many urban parks, solar canopies have emerged as a prominent feature. These structures, made up of solar panels, provide shade for visitors while generating electricity. Strategically placed in picnic areas or overlooking ponds, these canopies blend functionality with aesthetics.

Additionally, solar shelters equipped with charging ports offer visitors a chance to charge their electronic devices, from phones to e-bikes, harnessing the sun's power. This convenience not only promotes the use of the park but also subtly educates the public about solar energy's practicality

.

Interactive Solar Installations: Educating while Entertaining

In a bid to make solar energy more approachable and understood, many urban parks have introduced interactive solar installations. These could range from solar-driven kinetic sculptures to hands-on displays where visitors can see real-time data on energy generation and usage.

For instance, a solar-powered carousel, while delighting children, also serves as an educational tool. As families enjoy their time, they become more acquainted with the practical applications of solar energy, sowing seeds of curiosity and admiration for renewable sources.

Bridging Biodiversity with Solar

The biodiversity within urban parks is invaluable. Solar technology is being leveraged to support and nurture this biodiversity. Solar-powered sensors can monitor soil moisture levels, ensuring plants receive adequate water. In ponds, solar aerators help maintain oxygen levels, supporting aquatic life.

Furthermore, initiatives like solar bee houses have been introduced in some parks. These structures, powered by solar, provide a warm and safe environment for bees, crucial pollinators. Such endeavors underscore the harmony between technology and nature, where one supports and enriches the other.

Community Involvement: The Heart of Solar-Powered Parks

The success of solar integration in urban parks heavily relies on community involvement. From crowd-funded solar projects to workshops where locals can learn about solar applications, the community's engagement is pivotal.

Solar-driven community events, be it farmers' markets powered by the sun or nighttime open-air solar cinemas, further instill a sense of pride and ownership among the city dwellers. It reinforces the idea that sustainability is not a distant concept, but a tangible reality they are a part of.

Conclusion: Sunlit Paths to a Sustainable Future

As urbanization continues its relentless march, the integration of solar energy into parks and green spaces becomes not just desirable but essential. These spaces stand as testaments to a city's commitment to sustainability, offering citizens both solace and inspiration.

In the embrace of nature, with the sun's rays touching both the leaves and the solar panels, lies a vision of the future—a future where man, technology, and nature coexist in harmonious rhythm, each enriching the other, painting a picture of hope for generations to come.